APR 1 3

A SINGLE SKY

A SINGLE SKY

HOW AN INTERNATIONAL COMMUNITY FORGED THE SCIENCE OF
RADIO ASTRONOMY

DAVID P. D. MUNNS

THE MIT PRESS
CAMBRIDGE, MASSACHUSETTS
LONDON, ENGLAND

MIT Press books may be purchased at special quantity discounts for business or sales promotional use. For information, please email special_sales@mitpress.mit.edu or write to Special Sales Department, The MIT Press, 55 Hayward Street, Cambridge, MA 02142.

Set in Engravers Gothic and Bembo by Toppan Best-set Premedia Limited. Printed and bound in the United States of America.

Library of Congress Cataloging-in-Publication Data

Munns, David P. D., 1972–
A single sky : how an international community forged the science of radio astronomy / David P. D. Munns.
    p. cm.
Includes bibliographical references and index.
ISBN 978-0-262-01833-3 (hardcover : alk. paper)
1. Radio astronomy—International cooperation—History. I. Title.
QB475.A25M86   2013
522'.682—dc23
2012013561

10   9   8   7   6   5   4   3   2   1

to Reg Gardner—mentor, friend, and storyteller

At the very moment that humans discovered the scale of the universe and found that their most unconstrained fancies were in fact dwarfed by the true dimensions of even the Milky Way Galaxy, they took steps that ensured that their descendants would be unable to see the stars at all. For a million years humans had grown up with a personal daily knowledge of the vault of heaven. In the last few thousand years they began building and emigrating to the cities. In the last few decades, a major fraction of the human population had abandoned a rustic way of life. As technology developed and the cities were polluted, the nights became starless. New generations grew to maturity wholly ignorant of the sky that had transfixed their ancestors and that had stimulated the modern age of science and technology. Without even noticing, just as astronomy entered a golden age most people cut themselves off from the sky, a cosmic isolationism that ended only with the dawn of space exploration.

—Carl Sagan, *Contact*, 23–24

CONTENTS

ACKNOWLEDGMENTS

This book is about the history of a recent scientific community. The radio astronomy community is of interest because its members pursued knowledge of the heavens at a time when most scientists and engineers were concerned with the things of the earth—especially national defense and industrial development. The kernel of the book is an explanation of the interconnected growth of new instruments and the changes in pedagogy that can be traced to the formation of an international and interdisciplinary scientific community. The construction of new radio telescopes and new students forged the endeavor known as radio astronomy. But in turn, those instruments and students made the moral economy of the field, and ultimately produced new knowledge about radio stars, radio galaxies, and the age of the universe. In the first two decades of the Cold War, numerous practitioners and practices, instruments, and ideals transformed laboratories and radio receivers into observatories and telescopes. Those transformations occurred before the necessities of the Cold War world took hold of space sciences in the West. The radio astronomy community established a true scientific community in which the disciplinary knowledge of radio physics and that of optical astronomy were joined together, and in which national boundaries were crossed as easily as disciplinary boundaries. In constructing radio astronomy, the radio astronomers embraced an open and cooperative vision of science.

Like many of the characters in this book, I have learned much from my experiences in Australia, in the United States, and in the United Kingdom. Andrew Warwick became a close friend and a mentor and shaped my thinking about communities via pedagogy and students. Likewise, David Kaiser's overwhelmed physicist communities remain a powerful way of understanding scientists' experience and knowledge in the Cold War world, and his labor over successive drafts has been tremendously useful and appreciated. My graduate advisor, Bill Leslie, taught me to teach and to appreciate various types of communities. In Sydney, Nicolas Rasmussen introduced me to the history of instruments and the social construction of science. I owe a great debt to Bruce Hunt in Texas for his career guidance over the years. Allison Kavey forced me to become a better, and broader, historian and person. Pam Long took some first-year graduate students and filled them with wonderment and a drive to understand the production of knowledge. Frank Bongiorno—the consummate professional historian—serves as a continual, and much envied, inspiration.

Over the years, in classes and seminars, at conferences, and at the pub, many people have aided the realization of this book in myriad ways. I would like to thank Alan Chalmers, Suman Seth, Michael Shortland, and Ragbir Bhathal in Sydney, who got me started in the history of science; Robert Smith, Robert Kargon, Sharon Kingsland, Larry Principe, Bruce Hevly, Greg Downey, Scott Knowles, Josh Levens, Lloyd Ackert, Alexa Green, Hunter Heyck, Jesse Bump, Kathleen Crowther, Buhm Soon Park, Hyungsub Choi, Tom Lassman, Matt Wisnioski, and Sandy Gliboff in Baltimore, all of whom helped fashion my intellectual pursuits and my social community in Baltimore; David DeVorkin, Allan Needell, Paul Forman, and Michael Neufeld at the Air and Space Museum, for continuing to support the history of astronomy; Gail Schmitt; Patrick McCray; Peter Westwick; Susan Lindee; Rachel Ankeny; Amy Slaton, Kali Gross, Richard Dilworth, Joel Ostereich, Erik Rau, Gina Waters, and Don Stevens in Philadelphia, where I taught and learned a lot about audiences; and Rob Iliffe, Abigail Woods, Andrew Mendelsohn, Serafina Cuomo, David Edgerton, Hannah Gay, Graham Hollister-Short, Emily Mayhew, Catherine Jackson, Hermione Giffard, Max Stadler, Lesley Harris, Robert Bud, Hasok Chang, Jon Agar, Brain Balmer, Joe Cain, and Jane Gregory in London, for an intellectually wondrous few years. Finally, I thank Lord Robert Winston for challenging discussions and his passion for science. In addition, I must sincerely thank various unnamed editors and referees of many pieces of work whose detailed comments always served to improve my arguments and my expression.

The production team at the MIT Press deserves many thanks for helping me through the travails of the publishing business. Margy Avery was an enthusiastic supporter of the project from the outset, and especial thanks must go to her for moving the book into actuality. I also would like to thank that answerer of a thousand questions, her assistant Katie Persons, and the rest of the MIT Press.

Librarians and archivists are always important for any work. Among those I must thank are Edmund Rutlidge and his staff at the Australian Archives and Rodney Teakle and his staff at the CSIRO Archives. Some of the important documents were in the United States, and I thank the Rockefeller Archives Center in Sleepy Hollow, New York for assistance in locating them and funding the research. I would like to especially thank Professor Irwin Shapiro, Chair of the Harvard Center for Astrophysics, who kindly allowed me to inspect the records of the Harvard College Observatory in the Harvard College Archives. The Maurice A. Biot Fund and Shelly Erwin helped guide me through Caltech's Institute's Archives and made my stay at Caltech very enjoyable. The staff of Churchill College at the University of Cambridge and the staff of the Bibliotheek der Universiteit at Leiden were both fantastic. I thank the Huntington Library in Pasadena for financial support, and that library's wonderful staff—particularly Dan Lewis—for their assistance. The working and physical environment of the Huntington was idyllically conducive to research and writing, and my fellow scholars were lively, challenging; we lived the seduction of communities of scholars. Thanks too to the staff of the Johns Rylands Library in Manchester, Anna Mayer of the Royal Society, Rosanne Walker at the Adolph Basser Library of the Australian Academy of Science, the staffs of the Bancroft Library at the University of California at Berkeley, the Library of Congress, the Australian National Library, the library of the Cavendish Laboratory at Cambridge University, the Caltech library, the Weidner Library, the British Library, and the libraries of Johns Hopkins University, Drexel University, and Imperial College London.

Lastly, the many friendships that have been made along the way have, on so many ways, shaped my life and this work. To Matt Wisnioski and Cindy Rosenbaum, Andrew Warwick, Catherine Jackson, all the Zylmans, Richard and Katherine Windeyer, Lajos Bordas, Heather Rowlinson, Scott Knowles, Colin Milburn, Luis Campos, John Apperson and Sarah McAtee, Brad Oister, Diane and JJ, Patrick Griffin, Charlotte, Carrie, and Samantha, I appreciate all our time together. Likewise, there are the families that provided essential support to my international wanderings over the years: the Baradine relatives, the Hungarians, the Wisnioskis, the Philadelphia family, and of course my own family. My parents, Peter and Sue Munns, continue to lovingly supported my work and me, as do Trudeke and David MacKay, who, along with Max and Lillian, bring much joy into my life.

I extend my thanks for permission to republish the parts of this book that have appeared previously in "If We Build It, Who Will Come? Radio Astronomy and the Limitations of 'National' Laboratories in Cold War America" (*Historical Studies in the Physical and Biological Sciences* 34, 2003, no. 1: 95–113; © 2003 Regents of the University of California; published by the University of California Press).

In the 1995 film *The Englishman Who Went Up a Hill but Came Down a Mountain*, Hugh Grant portrays Reginald Anson, a floppy-haired youth recently returned from World War I. Working as an assistant with the Royal Ordnance Survey, Anson takes part in measuring the "first mountain in Wales," known to the locals as Ffynnon Garw. The film, which is based on real events, humorously exposes what is at stake in the history of science: the decision to measure, the process of measuring, and the result of having measured something are only the beginnings of an intricate social process through which people view and value their world. The audience sees the history of science played out as two men estimate Ffynnon Garw's height, first by foot and then by means of instruments with reference to other mountains whose heights, the villagers are disturbed to learn, also are measured only in relation to other mountains. Worse still, Anson's measure of the mountain—984 feet—reduces its stature to that of a "hill," since the British standard for a mountain requires 1,000 feet. The bitter disappointment this designation brings awakens the nationalism of the Welsh villagers. As they rally to confront the issue, one protagonist implores his neighbors: "This is a mountain, *our* mountain, and if it needs to be a thousand feet, then by God, let's make it a thousand feet." Subsequently, the audience laughs alongside the heroic villagers' struggle to add 20 feet to their mountain, all the while engaged in covert antics to keep the cartographers in the village long enough to demand reevaluation of Ffynnon Garw as a mountain.

I begin with this story because it encapsulates how a seemingly trivial scientific exercise can expose the very heart of people's identities and communities, and can—literally, in fact—move mountains. Social context lends substantive meaning to what we measure and to how we measure it. While the Royal Ordnance Survey sought to unify and standardize the topographical features of the British Empire, the villagers living under Ffynnon Garw regarded the measurement as an exercise in properly identifying what made them Welshmen and, pointedly, how geographically and culturally separated they were from Englishmen. Indeed, the whole process of the measurement exposed what the villagers considered most important. The narrator of the story explains it this way to his grandson: "Is it a hill, is it a Mountain? Perhaps it wouldn't matter anywhere else, but this is Wales. The Egyptians built pyramids, the Greeks built temples but we did none of that, because we had mountains. Yes, the Welsh were created by mountains: where the mountain starts, there starts Wales. If this isn't a mountain, . . . then Anson might just as well redraw the border and put us all in England, *God forbid*."[1]

Like the work of history, the work of science often mixes natural and social categories. The story of the measurement of a "Welsh" "mountain" serves as an analogy to the subjects of this book, namely the history of a new science and the social meaning of a new instrument of science. Like the measurement of a Welsh mountain by English cartographers, the measurement of the stars by means of radio waves challenged and changed the meanings of "astronomy" and of "telescopes." In the first two decades after World War II, "astronomy" became understood as a science that examined radio-wavelength *and* visual-wavelength emissions from stars and galaxies. In building a new meaning of astronomy, the radio astronomers pivotally redefined the idea of the astronomical telescope to include both giant radio antennas and optical telescopes. In effect, the radio astronomers changed the idea of astronomy as a science. For more than 3,000 years, astronomers had used light visible to the naked eye to study the working of the heavens. Technology had changed astronomy before, of course. After Galileo, astronomers used optical telescopes; later still, they also used spectroscopes and photographic plates. Quite recently, astronomers have successfully adapted charged-couple detectors, rockets for launching instruments into space, and computers to their science. All these developments are important, but they are all essentially additions to the optical telescope as Isaac Newton or William Herschel understood it. Radio was different.

After 1945, the technology of radio affected astronomers' vision of the heavens by fundamentally altering how they saw. The evidence of the science of astronomy was no longer constrained solely to visible light. Suddenly astronomers had access to another large segment of the electromagnetic spectrum at radio wavelengths to complement the visible wavelengths. Much of this book is about the historical process of learning to see by means of radio waves. The significance of radio to astronomy was sudden and spectacular. The eminent Dutch astronomer Jan Oort announced in his lectures that he considered the opening of the heavens by the radio telescope to be as revolutionary as Galileo's first observations with an optical telescope.[2] The radio telescope exposed vast swaths of the celestial heavens to investigation and permitted the new radio astronomers to peer through dust clouds and gases, which until the 1940s had limited even the visible part of the astronomical horizon. Because radio waves penetrate the atmosphere, as does visible light, radio astronomy didn't have to wait for reliable rockets with which to launch detectors into space. It is revealing of the entire nature of science that astronomy, one of the longest-studied and most coherent bodies of knowledge, had, before radio astronomy, been limited to half an order of magnitude of the electromagnetic spectrum—the visual range (approximately 450–800 nanometers). Trying to discover evidence of the shape of the galaxy or the laws of the universe under such limitations might be comparable to reading only one middle paragraph of a newspaper page, perhaps less, and expecting to know the day's news. Now, after more than 50 years of radio astronomy, our knowledge of the size and the structure of the cosmos has expanded as our vision has widened. Radio astronomy was, in fact, the first of a whole set of new astronomies. Modern astronomers now "see" not only in the radio range (approximately 1 millimeter–50 meters) but also in the x-ray range (approximately 0.004–10 nanometers), in the ultraviolet range (approximately 0.1 micrometer–350 nanometers), and in the infrared range (750 nanometers–1 millimeter). To paraphrase a noted writer of science fiction, it is not that the universe is more astonishing than we can imagine; it is that it is more astonishing than we can see with only our eyes.

How can we begin to grapple with these momentous changes that happened so rapidly to an ancient and venerable science? Clearly, the dramatic changes in optical astronomy since World War II offer a parallel.[3] As Robert Smith's study of the space telescope and Patrick McCray's survey of the new array of optical telescopes demonstrate, themes familiar to Cold War historians—the rise of big science, the military-industrial complex, new government patronage—altered astronomy significantly.[4] Likewise, the new instruments and new institutions of radio astronomy stemmed, in part, from exactly those contexts. Radio astronomy's instrument, the "radio telescope," was shaped by the wartime technology of radar and by the established standards of traditional optical telescopes. New radio telescopes led to "radio observatories," which required the support of a new scientific community. The moral economy of that community centered on the values of cooperative, open, international and interdisciplinary science. Those values were readily apparent in the recruitment and training of new disciples. The students of radio astronomy learned radio and optical techniques side by side, and shaped their telescopes and data to make the heavens recognizable to both optical astronomers and radio physicists.

Any history of a community weaves strands of seemingly disconnected people together to explain the forging of a larger moral economy that shaped their ideals and actions. The radio astronomers were conscious of their new community, were active in its making, and worked to transcend the limitations of the Cold War, especially national and disciplinary competition, secrecy, and subservience to the military-industrial complex.

Most of the events in the formation of the radio astronomy community occurred in the years 1944–1964. Those events took place simultaneously at the Australian Radiophysics Laboratory in Sydney, in the physics department at the University of Manchester, in the Cavendish Laboratory at Cambridge, at the Leiden Observatory in the Netherlands, at the Harvard College Observatory in Massachusetts, at the California Institute of Technology, and at the new National Radio Astronomy Observatory in West Virginia.

The radio astronomy community began with Edward "Taffy" Bowen becoming Chief of Australia's Radiophysics Laboratory, Alfred Charles Bernard Lovell reentering the dusty halls of the University of Manchester, Martin Ryle negotiating to stay at the Cavendish Laboratory in Cambridge, and Henk van de Hulst, in the newly liberated Netherlands, wondering whether he could detect galactic hydrogen at radio wavelengths.[5] Of those four, only van de Hulst was a recognizable astronomer in 1945. The others—Bowen, Lovell, and Ryle—had worked for years in the vast British radar effort.[6] Although two Americans had been the first to detect interstellar hydrogen, and although American observatories dominated the field of optical astronomy,[7] nearly every one of the founders of the radio astronomy community was from either Australia, Britain, or the Netherlands. Yet the community needed established optical astronomers as well as young radio physicists. Crucially, American astronomers— among them Bart Bok and Donald Menzel at Harvard and Jesse Greenstein at Caltech—embraced the new techniques of radio, brought the radio physicists into their observatories, and taught astronomy's new disciples about electronics and galactic structure. Only by considering all these institutions, disciplines, and people together can we understand how cosmic hiss, radio telescopes, and radio observatories all came to be considered "astronomy."

The individual efforts of Karl Jansky and Grote Reber before World War II, recently detailed by the radio astronomer and historian Woodruff Sullivan, were important and interesting precursors to the emergence of radio astronomy. Independently, and rather enigmatically, Jansky established the Milky Way galaxy as a generalized source of cosmic noise (or, as he called it, hiss). Meanwhile, Reber built a prototype parabolic dish-style radio telescope in his back yard in Illinois. As Sullivan concedes, the work of those two men didn't spur a larger postwar effort in radio astronomy, as professional astronomers' interest quickly waned after only limited initial enthusiasm.[8] Rather, the effort was spurred by "sharp rivalries" between groups, as David Edge and Michael Mulkay noted in their sociological study of the history of radio astronomy, *Astronomy Transformed*. Sullivan does, however, concede an important exception in the history of radio astronomy to any general conception of the inherently competitive nature of science. Sullivan argues that in the "unusual" case of the discovery of the 21-centimeter radio wavelength line of neutral hydrogen—a discovery that made much of the charting of the structure of the Milky Way galaxy possible—cooperative "close ties" integrated diverse research groups, and cooperation, not competition, led to progress.[9]

A Single Sky takes issue with the idea that recent science has been driven by competition. The radio astronomers understood science as an open, inclusive, international, interdisciplinary process, and their community succeeded because of cooperation. In 1954, the Australian radio astronomer Joseph Lade Pawsey, speaking as president of the International Astronomical Union Commission, said "Radio astronomy, if it is to develop properly, must depend on a blending of radio invention and astronomical insight."[10] Pawsey's claim that radio astronomy would have to "blend" insight and invention may be the most succinct statement of the radio astronomers' view of the creation of a community style of science—a style that they considered entirely "proper." The culture of the radio astronomers was a culture of interdisciplinary and international integration and cooperation.[11] An emphasis on cooperative community places this book at odds with the philosopher of science Thomas Kuhn, who argued that "competition between segments of the scientific community is the only historical process that ever actually results in the rejection of one previously accepted theory or in the adoption of another."[12] In the case of the emergence of radio astronomy as a science, an emphasis on competition among schools, nations, theories, or even technologies would incorrectly characterize an altogether cooperative process of community building. Indeed, much of this book is devoted to explicating the concept of community in one recent science. Instead of a fractious world of science, the radio astronomers saw a single sky, unifying both nations and disciplines.

Unlikely as it sounds, disciplinary and international unity began during wartime. As everyone who has written on radio astronomy has noted, the culture of wartime radar research and development deeply impressed the future radio astronomers. The famous science storyteller Arthur C. Clarke argued as early as August of 1945 that radar (RAdio Detection And Ranging) had been the miracle weapon of World War II. Utilizing highly sensitive radio sets receiving weak reflections from ships and

FIGURE I. I
Sir Martin Ryle at the command console of the Mullard Radio Astronomy Observatory at
Cambridge. Reproduced with permission of the Astrophysics Department of the Cavendish
Laboratory of Cambridge University.

aircraft, radar had figured in the Battle of Britain in 1940 and in the defeats of German
and Japanese naval forces in 1944 and 1945. Looking to the future, Clarke envisioned
almost limitless potential rewards from radar and every other wartime technology. The
war had shown how much power science could have when dedicated to killing
people; imagine its power, Clarke asked rhetorically, if it it were to be utilized to
uncover fundamental truths about the universe. He added that astronomers, possessing
little experience with electronics, would have to persuade people to build electronic
devices for them.[13]

Whereas radio physicists became radio astronomers, optical astronomers become
all-wave astronomers, using visible light and radio waves together. In 1952 the Harvard
College Observatory hired a physicist to build a radio telescope, and in 1955 Caltech

and Mount Wilson recruited an Australian radio physicist to build one. When astronomers made arguments to justify new expenditures on a radio receiver and a physicist,
or when a physics department made an argument to pursue astronomy, both groups—
astronomers and radar physicists—changed how knowledge about the heavens was
collected and what that knowledge meant.

By the 1960s, giant radio telescopes in Australia, Britain, the Netherlands, and the
United States were looking at the universe in new ways. The radio astronomers would
reveal the spiral structure of our own galaxy, discover an entirely unexpected and
strange class of ultra-intense objects called quasars, and measure the universe's residual
background temperature of 3 degrees Kelvin left over from a "big bang." Radio
astronomy would change our ideas of how the universe looked, how we learned about
it, and even our idea of how old it was. It would provide evidence for the great
cosmological debates between the "steady state" and the "big bang." And in July of
1969, after several scheduling changes, Australia's Parkes radio telescope would receive
the television images of the first moonwalk and re-transmit them to the world.

The first major axis of this book is a history of material culture. In the years after
1945, regardless of Clarke's optimism, radio techniques and radio equipment presented
profound challenges to optical astronomers, who found early radio observations
difficult to integrate into their scientific practice, just as in the sixteenth century
naked-eye astronomers had found it difficult to master the telescope.[14] Traditional
astronomers, now suddenly revealed to have been constrained to the optical range,
wondered how they could make use of radio telescopes and "super-heterodyne
receivers" or how they might read the information generated on chart recorders.
Radio astronomy didn't merely extend sight; it became an entirely different sense of
vision. Much of the formation of the community is nicely encapsulated in the
struggles of optical astronomers and radio physicists to incorporate photographic plates
and visual spectra with radio graphs. The overlapping pictures looked more like
weather forecasts than like star fields. Community became necessary because radio
supplied information about the heavens that had to be translated into the language
of the astronomers.

Chapters 4 and 5 explain the development of a new generation of "giant radio
telescopes," most of which looked like radar dishes rather than telescopes. As the
astronomical community changed, its conception of a telescope changed to include
radio telescopes; as the form and function of a telescope changed, so too did the
astronomical community that lent legitimacy to new instruments and practices. In
radio astronomy and in optical astronomy, the idea of astronomy itself changed, both

FIGURE 1.2
The new radio vision of the heavens. Above: a chart showing radio intensities at three wavelengths (1,390, 408, and 85 megacycles per second) from the galactic plane between 10° and 340°. Below: a radio map of Cygnus X, the contours showing the change in radio intensities across the object. From "Progress Report no. 3," box 59, file 1, Jodrell Bank Archives. Reproduced courtesy of Jodrell Bank and the University of Manchester.

FIGURE I.3
The first "Mills Cross" interferometer, where signals from the two arms near Sydney were combined, September 14, 1954. Lee DuBridge Papers, Folder 35.2. Reproduced courtesy of California Institute of Technology Archives.

because of new technologies and because of the new social context of Cold War–era science. As Sharon Traweek reminds us, "like the environments we build, the artifacts we make remind us of who we are, how we are expected to act, and what we value."[15] In any number of cases in the social history of technology—for example, the victory of swords over guns in Japan, the victory of AC over DC electricity in New Jersey, the victory of the internal-combustion car over the electric car in the United States, or the victory of the electric refrigerator over the gas-powered refrigerator—we learn that social meanings shape the adoption and the use of technologies.[16] Radio telescopes didn't have to look like telescopes or operate like observatories. Yet, as we shall see, the radio astronomers chose to make laboratories into observatories, antennas into

FIGURE 1.4

Harvard University's 60-foot radio telescope. Cover of *Sky and Telescope*, July 1956. Widener Library, HUF 165.881. Reproduced courtesy of Harvard College Library.

telescopes, and noise into vision. In other words, we can read the values of the community through the telescopes they chose to build and use.

Like the Harvard Observatory's radio telescope, each of the four major first-generation radio telescopes, built up to about 1961, was "parabolic" in design. Yet not a decade later Martin Ryle would be awarded a Nobel Prize for the development of aperture synthesis (interferometric) radio telescopes. As in many other cases, what later came to be regarded as the best choice was not seen as the best choice at the time. In the creation of radio astronomy and in the formation of the radio astronomy community, the parabolic design fulfilled the ambition of all participants to make radio physicists into radio astronomers. It was a social choice, guided by an ambition to unify disparate elements into a coherent community.

Though the choice between parabolic and interferometric designs for radio telescopes may have been the major decision for the new radio astronomy community, it was far from the only decision. In 1956, the Harvard physicist Edward Purcell, already a Nobel laureate in physics, speculated that the radio telescope of the future "may . . . want a small cryogenic laboratory mounted out there on the end of it."[17] He asked whether the new technology of radio, not to mention rockets, electronics, and nuclear physics, had rendered the traditional optical telescope obsolete. The technological reshaping of astronomy after 1945 ultimately turned on what the word 'astronomy' meant to those who called themselves astronomers. Purcell asked pointedly why they built an observatory and not a laboratory, and why they didn't simply redraw the disciplinary lines of science and call everyone a physicist—*God forbid*.

Instrumental developments alone don't tell the whole story of astronomy, or indeed of any science. The recruitment and training of students is the book's second major axis.

In large part, the new disciples of radio astronomy existed in the context, but also in the shadow, of Cold War–era physics. Physics has provided the standard metanarrative for science during the Cold War. With its newfound cultural cachet, students flooded graduate programs in physics. Simultaneous with the nearly manic pleas for more and more physicists that came from the government, from physicists themselves, from industry, and from the American Institute of Physics, the postwar production of physicists became associated with national security. In the United States, the number of physics graduates doubled between 1949 and 1958, while 50 percent more schools offered PhD programs, taking advantage of the National Science Foundation's substantial support for training.[18] Some subfields of physics benefited even more. The historian Spencer Weart once pointed out that the number of PhDs in solid-state physics increased by a factor of 5.[19] On the eve of Sputnik, physics was the specialization of 103 (87 percent) of the 118 graduate students in Caltech's Division of Physics, Mathematics, and Astronomy, while the proportion of undergraduate physics majors had gone from 20 percent in the late 1940s to 33 percent by 1956–57. And student numbers, at least at Caltech, translated directly into increased staff appointments in physics and mathematics.[20] Clearly the fortunes of astronomy had waned.

After Sputnik, the nation's attention turned toward space science, and consequently astronomy grew in ways akin to other sciences under the umbrella of the military-industrial complex.[21] Until then, however, astronomy struggled. The number of students in a field is an apt marker of that field's prestige and power. Quite simply, students didn't flow into astronomy after World War II as they did into physics. Between 1940 and 1958 the number of institutions offering the astronomy PhD in the United States decreased from 14 to 11, while their production of doctorates increased only meagerly, from 98 in the period 1940–1949 to 128 in the period 1950–1958—an especially damning figure in view of the almost complete suspension of doctoral work in astronomy during World War II.[22] The dearth of new recruits shocked the astronomers. Postwar expectations envisioned bigger instruments, larger budgets, and more disciples. In California, Ira Bowen, the new director of the Mount Wilson and Palomar observatories, and Jesse Greenstein, the head of Caltech's new graduate program, anticipated "many more applicants than we can handle without any attempt at formally advertising the school" in 1948.[23] Their optimism rested on what they saw as the formidable unification of the mountains' observational equipment (which now included the world's foremost optical telescope, the 200-inch Palomar reflector) with Caltech's new graduate school in astronomy. But students would turn out to be fickle. Only four years later, Bowen would lament to Vannevar Bush, the architect of postwar science, the "failure to get men . . . into astronomy caused [at least partly] by the glamour of nuclear physics and electronics."[24]

Bowen's lamentation over the dearth of new disciples made it plain that the recruitment and training of students—in other words, pedagogy—formed a vital context for the history of the new astronomy community. My approach to understanding the training of scientists is deeply indebted to Andrew Warwick's study of the world of the nineteenth-century Cambridge Mathematical Tripos. *Masters of Theory* illuminates a culture of manly behavior, intellectual coaching, and the passage through the examination. Warwick demonstrates how the intellectual and social community of Cambridge delimited the scientific problems to be solved, formed the basis for their solution, and occasionally produced novel solutions in the examination room. Warwick pioneered the study of the material culture of instruction. He not only charted the familiar twin loci of institution and discipline, Cambridge and Mathematical Physics; he also explained pedagogy as culture. By uncovering a culture of coaching, problems, exams, hard work, sport, teamwork, masculinity, and individual glory for the Senior Wrangler (and for the winner of the wooden spoon), he revealed how scientific communities defined the problems of normal science by training new disciples.[25]

Historians of astronomy generally have focused on the familiar contexts of changing patronage and changing instruments, and not on pedagogy.[26] Drawing on work by Kathryn Olesko, I agree that we might profit from "de-emphasizing the products of science, ideas and theories, in favor of exploring the labor of science."[27] Labor, especially the labor of students, explains the form of the new scientific community of radio astronomy. A focus on pedagogy permits access to the cultures that bind scientific communities together to make knowledge translatable, movable, and useful. Apprenticeship in modern science's graduate schools significantly imparts the community's social structure, mores, and ideals, and may also serve as sites of resistance and change.[28] Training, as Michel Foucault noted, is about both education and control.[29] Recruitment and training shape the membership of a community through learning, adopting, or creating identities through commonalties, particularly of language. (George Chauncey would call the latter "code words."[30]) Commonalities are identity keys that members of any community use with each other as well as act to distinguish themselves from others. Since keys have to be learned, or made, a scientific apprenticeship becomes the foundational feedback moment of expression, representation, and identity.

Recruiting and training new disciples has been part of science's reward structure since the emergence of professional science in the nineteenth century, if not longer. A disciple remains one of any scientist's most important products, not to mention an expression of the power of a scientific program, school, or institution.[31] The ability to attract students is a marker of status in science. Even that epitome of the lone scientist Stephen Hawking, though emphasizing "the achievements of Newton, Babbage, [and] Dirac" as researchers to judge the success of the mathematical sciences, also noted that Cambridge, as the site of the Lucasian Chair, "attracts leading mathematical scientists," "international visitor[s]," and "graduate students" and thus endows the institution and chair with "excellence."[32] In a more familiar case, the physicists' culture of abundance shaped the training of physics' disciples during the Cold War. As David Kaiser noted, the rise of Feynman diagrams became a particular pedagogical strategy for producing mechanical physicists in the face of a veritable flood of students at a time when the presence of too many new faces in too many new departments was breaking down the traditional and much-romanticized ways of personal mentorship.[33]

The case of early Cold War astronomy shows, in effect, the exact opposite. It was a paucity of students that caused many changes in the very conception of astronomy, from its instruments to its objects of study. Ira Bowen's understanding of the moral capital of students underwrote his lamentation about Palomar astronomy's lack of disciples in the face of institutional and instrumental dominance. In both the creation of new instruments and the creation of new curricula, the act of apprenticeship emerged as a significant feature of the new radio astronomy community. The new community was a result of signature moments for the radio astronomers. In 1951, when Martin Ryle held an "open day" at Cambridge for potential new graduate students, not a single person expressed interest. Not four years after the world's largest telescope saw first light, Caltech's recruitment campaign was declared a failure.

The astronomers expanded their community to include radio as one response to the dearth of students. As newer, more glamorous sciences drew resources (especially students) away from astronomy, the head of Caltech's graduate program, Jesse Greenstein, argued that astronomy's continued existence depended on attracting "brilliant young men interested in pure science" with offers that were competitive with the "fields of industry, engineering, and physics where the ultimate salaries are high and financial security greater."[34] By the early 1950s, Greenstein's patron of choice was the National Science Foundation. Only the NSF could offer the bigger financial lures that were needed in order to attract new students. Greenstein's ambition, a "doubling or tripling" of astronomy's 1 percent allocation of the NSF budget, reveals the scale of the battle the astronomers fought. This was not like the physicists' fighting over the next multi-million-dollar cyclotron. Here, like Oliver Twist asking for more, was the head of the "astronomical center of the world" daring to ask for 2 percent of the NSF's allocation instead of only 1 percent. Seeking to get even that much, the optical and radio astronomers forged a new culture of "research and education." In contrast to the glamorous scientific culture in which many young people entered an education transformed to suit and service the needs of industry and the Atomic Energy Commission, looking at the output of a radio receiver and calling oneself an astronomer was essentially a struggle for the very soul of science.

The book's third thematic axis concerns the reality and rhetoric scientific communities transcending boundaries, especially disciplinary and national boundaries. Particularly for the Cold War era, the history of American science remains the dominant narrative.[35] Singular national or disciplinary narratives have skewed our entire picture of the period, and have imbalanced our explanations of scientist's choices and the style of their knowledge.[36] Alongside the multidisciplinary nature of large-scale scientific endeavors, the radio astronomers forged an international community through a nearly continuous trade in expertise among Australia, Great Britain, the Netherlands, and the United States.

Global histories of communities are challenging because many studies of community rely on a fixed geographical feature. Studies of changing conceptions of gay identity, for instance, have relied on the Castro district of San Francisco. Similarly, studies of changing notions of environmental consciousness have relied on sand dunes outside Chicago, and studies of the changing ideas of small-town community life have relied on a hamlet called Sugar Creek.[37] The strength of this approach is that when people move into, out of, or through established localities, they reveal the changing nature of their identities and communities. After World War II, scientists moved into, out of, and often through scientific disciplines, and thus a community study is well suited to understanding that new social dynamic in science.

In astronomy, the technical inclusion of wavelengths other than those of light paralleled the social transformation of including all scientists. A Soviet astronomer, V. A. Ambartsumian, expressed belief that "contemporary astronomy has come close to becoming all-wave astronomy." In a wonderful example of the social order constructing the order of nature, Ambartsumian's comment came at a 1973 conference on communication with extraterrestrial intelligence at which all-wavelength astronomers sought not only to establish contact with extraterrestrials but also to solve the problems of "communication [among] nations." In other words, working on extraterrestrial communication would help astronomers to deal with one of the fundamental barriers of the Cold War world: the Iron Curtain.[38] If astronomers could figure out how to talk to extraterrestrials, they could figure out how Americans might talk with Russians, and vice versa.

Crucially, the intellectual and social transformation of astronomy into all-wave international astronomy took place via the social construction of the scientific community itself. The community emerged as a process of joining scientists across boundaries of all kinds. As early as 1950, the leading American astronomers insisted that the radio physicists and astronomers were "not close enough." As the Caltech astronomer Jesse Greenstein explained, "what contact there has been" had been "through the personal interest of a few astronomers," but it was "perhaps not enough" to bridge the divide. Along with Donald Menzel and Fred Whipple of the Harvard College Observatory, Greenstein sought more symposia and more "visits of radio-observing personnel to various major observatories."[39] Revealingly, the Dutch radio astronomer

Henk van de Hulst might have declined an offer to visit Palomar Mountain in California in 1951, but only because he felt "saturated with 'informal communications' and 'astronomical gossip' in general."[40] Every major early figure in radio astronomy actively participated in its communitarian creation. The Australian radio physicist Joseph Pawsey spent more than 12 months in the United States and Britain in 1951 and 1952. The Mount Wilson astronomer Rudolph Minkowski visited Australian radio astronomers for nine weeks in early 1956, then spent a week at Cambridge, a week at Manchester, and a week at Leiden.[41] In 1959, Gart Westerhout of the Leiden Observatory toured all the major radio astronomy sites in the United States in the course of several months.[42] Edward Bowen, chief of the Australian Radiophysics Laboratory, traveled to Britain and the United States from Australia at least a dozen times in the course of the 1950s. In addition, more permanent moves cemented the bonds of community in place. Bart Bok of Harvard, a galactic astronomer and the author of one of the first courses on radio astronomy, moved to Australia in 1957. Robert Hanbury Brown of Manchester, Bernard Lovell's right-hand man, followed Bok in 1964. Edward Bowen's protégé John Bolton moved to Caltech in 1955 to head that school's radio astronomy program, and the Deputy Chief of the Australian radio astronomy effort, Joe Pawsey, was appointed the Director of the American National Radio Astronomy Observatory in 1961. And astronomers, radio astronomers, and radio physicists would meet regularly at events held by the International Scientific Radio Union and by the International Astronomical Union. With these visits, exchanges, meetings, and correspondences, scientists crossed boundaries of all kinds, most immediately the boundaries between nations. Underpinning the formation of the radio astronomy community was an ideology of internationalism, which has long existed in science, of course, because scientists' worldview emphasizes the universality of genuine knowledge.[43] I agree with Ann Johnson that when seeking to understand how communities both create and legitimize new knowledge and social arrangements we must not "privilege either the social arrangements or the content and practices of science" but should appreciate the extent to which they are intertwined.[44] The new radio astronomy community built a material and intellectual culture and pedagogy, as well as an international and interdisciplinary community, but significantly the community built the culture and the culture reinforced the community. New radio telescopes and new curricula became sites of open cooperation between astronomers and radio physicists, between Australians and Americans, between graduate students and researchers, and between instrument makers and theoreticians.

To understand the radio astronomers' desire to reshape the social values of science, we must consider that crucible of Cold War scientific practice, World War II. That global conflict saw unprecedented scientific and technological progress. Every available scientific resource and scientific field was utilized, with chemists working on atomic bombs and zoologists on a "bat bomb."[45] For years afterward, the list of scientific and technological advances spurred by the war would turn scientists (even the German Wernher von Braun) into heroes.[46] Proximity fuzes, air traffic control, IFF (identification friend or foe), airborne radar, and target location were hailed as crucial to victory. Penicillin's industrialization is now the stuff of legend.[47] Scientists saw "interdisciplinary communication and collaboration" as normal, and multi-disciplinary teams were commonplace.[48] And for years after the war, radio physicists continued to argue that radar, not the atomic bomb, provided the standard example for supporting fundamental research to miraculously produce "specialized military weapons" as well as the "means of safe flight and control of aircraft and rockets."[49]

A single bomb eclipsed the radio physicists' contributions to victory. After the news of the destruction of Hiroshima, no one could get enough copies of the superficial Smyth Report (*Atomic Energy for Military Purposes*), which lionized the physicists' story of the making of the atomic bomb, whereas the 28-volume technical history of MIT's Radiation Laboratory, useful in innumerable ways, disappeared from the general public's mind. The nuclear physicists, not radar's physicists, emerged from the war adorned with laurels. Nicely synthesizing a host of biographies and autobiographies, David Kaiser noted how "physics after the war became a ticket to see the world at lavish international conferences and European summer schools, being feted all the while as globe-trotting heroes at home and abroad."[50] Nuclear physics became the glamour field of the era, and historical narratives of Cold War science have, by and large, ensured that "American physics has dominated the historians' postwar landscape."[51] Physics, connected to weapons and to industrial power, soared everywhere. The programs of Stanford, MIT, and Berkeley, models of Cold War academic "steeple building,"[52] suited the military's needs. Military patronage and service to the state swept aside the traditional roles of the universities. Frederick Terman at Stanford, for example, advocated not wasting time on undergraduate programs, but rather focusing all energies on graduate departments in areas people cared about (that is, paid for).[53] In Australia, Harry Messels' physics department at Sydney University and Mark Oliphant's new cyclotron within his Research School for Physical Sciences at the Australian National University received a great deal of attention from the postwar government. The British government, entirely without American support, built an atomic bomb and successfully tested it in Australia in 1952.

Everywhere, as the ecologist Eugene Odum (who worked closely with the American Atomic Energy Commission for years) noted, "atomic energy remained primarily a military technology."[54] The bomb itself took on a "holy status in the American psyche," and the military-industrial complex became a new "priesthood."[55] Critics of the military-industrial complex (including President Dwight D. Eisenhower, who coined the term) spoke bravely. So too did the father of radar, Sir Robert Watson-Watt. Having attended the third Pugwash Conference in Vienna, he harshly condemned any nationalist imperative toward further development or testing of nuclear weapons in the cause of peace or defense. The fusion bomb rendered national distinctions mute: "[W]e are, quite clearly, no longer masters in our own national houses. [The] H-bomb has shattered our ability to determine our own national future within our own frontiers."[56] Though Watson-Watt himself had been an inventor of military technologies during World War II, by the dawn of the Space Age he too had become disillusioned with justifications of national security for nuclear weapons. One could not escape the essential tension that the very existence of those devices undermined science and the nations that possessed them, since they required international controls and limitations on testing in order to provide security.

We still know very little about the huge American weapons program; as Peter Galison once noted, we "are living in a modest information booth facing outwards, our unseeing backs to a vast and classified empire we barely know."[57] The glimpses behind the atomic curtain we have been afforded, however, tell a chilling story. The Atomic Energy Commission's scientific program "tested" 1,030 atomic weapons from 1945 to 1992—921 of them in the United States.[58] Weapons development took place within the new National Laboratory system, subverting the very basis of open, free, communal exchange of knowledge. As Peter Westwick noted, in contrast with the open universities, the "geographic isolation" of the National Labs' three centers of research "precluded close collaboration." Though the boundaries of secrecy were eventually relaxed within the system by offering special "facsimile" conferences and publications to the labs' scientists, the facsimiles only served to further "gate" the laboratory communities "even more tightly." Similarly, although there were facilities for international visitors and for students, "classified reactor work interfered with access to Argonne and Oak Ridge." In fact, the trustees avowedly denied that Brookhaven National Laboratory would "assume the functions of a university."[59] All the while, universities scrambled to acquire facilities more like the National Laboratories. By the middle of the Cold War, Caltech's Jet Propulsion Laboratory, Johns Hopkins' Applied Physics Laboratory, Columbia's Radiation Laboratory, and MIT's Lincoln Laboratory employed much of the nation's scientific talent.[60]

Tensions over openness and international cooperation existed in a complex-style science such as nuclear engineering as much as in a community-style science such as radio astronomy. Between 1957 and 1974, for example, Project Plowshare, the civilian component of the weapons tests, spent $770 million on plans to construct canals and harbors via nuclear explosions, outspending all radio astronomy efforts combined by an order of magnitude.[61] Yet behind all the money and effort spent on safeguarding national security via a weapons program we still recognize the values of the scientific community. It is revealing, for instance, that Project Plowshare aimed to turn weapons to useful purposes for "all mankind, Russians as well as Americans," such as a new Panama Canal or a dam across the Bering Strait.[62] One immediately notes that "all mankind" evidently consisted of two camps, the Russians and the Americans, and it is precisely this narrow bipolar understanding of the Cold War that is being valuably re-addressed in recent work.[63] Still, we have to acknowledge that Project Plowshare, though lavishly funded and though invariably supported by local development agencies, never managed to conduct any tests outside military locations.

In other words, Project Plowshare's backers and scientists *tried* to convert a specific disciplinary and nationalistic endeavor into a practice of science more akin to radio astronomy; they wanted to establish international networks and gather multidisciplinary and cooperative teams of researchers to attack common research problems. In the early days of the Cold War, "community living" and "growing 'roots' in the community" undoubtedly appealed to young physicists who chose to work in the national laboratories or even in industry.[64] As Kaiser has emphasized, many younger scientists didn't consider the new Cold War laboratory conditions of secrecy, security, and working for the military either restrictive or burdensome.[65] Their rational and self-interested choices horrified their mentors. To physicists of the previous generation, whose romantic memories lingered upon walking and skiing with Werner Heisenberg and Niels Bohr and riding through New Mexico with J. Robert Oppenheimer, the suburbs must have seemed far indeed from the cloistered communities of Göttingen, Cambridge, and Copenhagen.[66]

The radio astronomers' new identity and community serves as a counterpoint to a metanarrative that regards science as having been firmly yoked to the whims of the military-industrial complex during the Cold War. In the first two decades of radio astronomy, most radio astronomers would justify their pursuit of pure knowledge over practical application as a return to the practice of ordinary science; it was the large-scale continuation of essentially wartime technical development that was exceptional. As early as 1944, Bernard Lovell, then working at Britain's Telecommunications Research Establishment on radar aids to guide heavy bombers to their targets, argued to his superiors that Britain must "take [50 people] away from this guarded enclosure and re-establish the facility of thinking."[67] Similarly, as Australia's giant radio telescope sought increased funding in 1955, its guiding visionary, Edward Bowen, revealingly commented that "even in research circles there has been a disappointing tendency to say that sheep are more important and that radio astronomy is all right for other countries."[68] In the United States, Merle Tuve, leader of the proximity fuze project during the war and later head of the Department of Terrestrial Magnetism at the Carnegie Institution in Washington and of that institution's small radio astronomy effort, declared in the mid 1950s that radio astronomy was "a study of the heavens[,] not just glorified electronics."[69] In the Cold War era, when science served nations both economically and militarily, the radio astronomers' successful pursuit of fundamental research and larger instruments outside of any single nation's military-industrial complex became a case in point of the widespread notion that even the most esoteric scientific investigations "constantly yields practical innovation of the greatest importance in our daily lives."[70] In short, the community constantly advertised that the international science of radio astronomy evinced a legitimizing case for the national patronage of fundamental research.

Indeed, radio astronomy became almost a "poster child" for the resurgent notion of science for its own sake. In his 1954 presidential address to the British Association of the Advancement of Science, Sir Edward Appleton argued that postwar Britain placed too much emphasis on the "applications of science in the practical life of our country." Appleton presented a vision of science "pursued for its own sake"; there was value in science's ability to "enlarge men's horizons and invest the world with deeper significance."[71] Evoking the experience of Edmund Hillary and Tenzing Norgay on Mount Everest, Appleton implored the audience to support scientific work that seemed to have little practical purpose. He concluded with a singular example of the kind of pure scientific inquiry that should be supported for no practical purpose: "The radio telescope has . . . shown itself to be an important adjunct to the world's greatest optical telescope."[72] Not only did the new working relationship between radio and optical telescopes forge the new international and transdisciplinary community of radio astronomers; it also symbolized scientists' expectations of turning wartime technology into a broader horizon for mankind.

The radio astronomers successfully gained their anticipated international community as well as their giant instruments of science without seemingly necessary concessions to expectations of nationalism or secrecy. The new American National Radio Astronomy Observatory, for example, supported by the federal government and specifically modeled on the "national" nuclear laboratories, eventually refused a demand that only Americans be considered for the position of director. Nationalist voices failed in that case because the radio astronomers insisted that they existed in an international community and decisions about scientific expertise based on national identity were facile. In 1961, a panel of astronomers representing the American National Science Foundation recruited the Australian radio astronomer Joseph Pawsey to direct the facility. Even at a "national" facility, the alternative social organization of the community prevailed. With the battle between centralized nationally focused laboratories and diffuse, university-oriented, and international observatories won, radio astronomy became one of only a few sciences to challenge the physicists' model of scientific practice. The international and interdisciplinary corrective case of the radio astronomers enables us to recognize that, even in the world of the high-energy physicists perpetually in the service of the state, the *expectations* of "normal" science were those of community, universality, and disinterest. In practice, of course, the demands of the Cold War state confined much of science and technology to particular locations and severely limited access to knowledge. Complex-style science perpetuated the guarded enclosures that the future radio astronomer Bernard Lovell witnessed during World War II—enclosures that, Lovell said, eliminated "thinking."

The radio astronomers believed "community" to be the measure and meaning of their scientific activities and their new vision of the heavens. Robert Putnam's classic study of community, *Bowling Alone*, emphasizes the constant struggle, especially in the United States, between "community" and "individualism."[73] The history and the sociology of science too have long recognized the tension that exists between individual credit for a discovery and any particular scientist's working life within a community.[74] At Manchester, Bernard Lovell strove tirelessly to become "part of the astronomical community" and consciously shaped his old radar habits into a "new astronomical technique."[75] The radio astronomers' community stood in contrast to older, narrow notions of disciplinary specialties. The French radio physicist Marius Laffineur, in his 1957 presidential address to the radio astronomy commission of the International Radio Science Union (URSI), spoke of a new spirit of scientific and social interchange between physics and astronomy, between technology and theory, and between a national and an international community. Owing to the power of transdisciplinary science, Laffineur said, "the radio engineers who founded this new branch of astronomy have in many cases become expert astronomers: on the other hand, astronomers have been quick to recognize its importance and have assimilated our techniques with immense possibilities."[76] The new radio astronomers thus constantly evoked and invoked the ideal values of scientific community: the assimilation of specialties, the integration of new sources of expertise, and, above all, cooperation to make new knowledge.

But before rushing toward some notion of an edenic open community, we must acknowledge that, for scientists and those who seek to understand science, there is quite an investment in maintaining disciplinary boundaries. The nineteenth-century-built edifice of the scientific discipline precipitated a major cultural transformation of science, a transformation still powerfully evident in most pedagogical regimes and publication outlets. As Michael Aaron Dennis notes, "the discipline, not the university, became the institutional framework in which the scientist viewed himself."[77] Half a century later, Atsushi Akera argues, the electronic computer became "an ideal artifact" through which to study the changes wrought to American institutions, since they "straddled the very [military and commercial] institutional boundaries" of Cold War research.[78] By the middle of the twentieth century, the discipline, the artifact, and the institution defined and defended the boundaries of knowledge and permitted only limited exchanges, all later reinforced by a military-industrial regime that employed a large number of scientists and engineers and prioritized compartmentalization and secrecy.

In contrast, historians of science have used the category "community" only weakly, generally only to denote any loose assemblage in the process of consolidation and organization before neat disciplinary models might take hold.[79] The philosopher of science Thomas Kuhn shifted the defenders of paradigms from a community to the discipline in his postscript to the second edition of *The Structure of Scientific Revolutions*.[80] In recent science studies, it is disciplines that formally resemble, and operate like, institutionalized and professionalized paradigms. Ann Johnson's framework of technical knowledge communities in nanotechnology, for example, ends familiarly with disciplinary struggles between departments seeking funding.[81] It continues the trope, familiar from the nineteenth century, according to which disciplinary science ensured standards and the comparability of results by socially dividing and conquering knowledge, thus permitting the rapid growth of science.[82] Then, in the twentieth century, a number of important studies showed that scientists, upon encountering new areas of knowledge and especially new instruments, regularly organized themselves into disciplines, and spent considerable effort advertising their disciplinary status. Genetics coalesced around *Drosophila* and molecular biology around the electron microscope, while geographical engineering became a discipline that used nuclear weapons for earthmoving.[83] Likewise, in solar system astronomy (a science allied to radio astronomy), Ronald Doel saw disciplinary change explicitly caused by the rise of "specialized research instruments" and the "proliferation of patrons" in the years after Sputnik, inserting a national and disciplinary astronomy into the established narrative of Cold War–era science.[84] Edge and Mulkay's classic study of radio astronomy in Britain unashamedly claimed that radio astronomy became a new discipline with the introduction of a new instrument.[85] They argued that the normalized role of any new large instrument, such as Bernard Lovell's Jodrell Bank radio telescope, was a major factor in the expansion of discipline-based knowledge. But Edge and Mulkay could not remove one thorn in their easy reproduction of discipline-based knowledge: Radio astronomy in Britain never really adopted some of the classic expressions of a discipline, particularly lacking a coherent publication outlet.[86]

In contrast, *A Single Sky* resists the compartmentalization of recent knowledge communities, such as radio astronomy, into *ad hoc* disciplinary boxes. An alternative important thread in the history of recent science has been the denial of the validity of ever-finer disciplinary divisions. No less a figure than Vannevar Bush said that science would suffer the fate of the Tower of Babel if knowledge growth continued to be "divergent and fragmentary": "Whole new sciences and branches of engineering appear, with their specialized societies and journals. Intensely progressive gatherings of research workers develop their own jargon, unintelligible except to the initiated, heightening the barriers which separate their work from the main stream of progress."[87] After people couldn't speak to each other any more, Bush went on, continuing the analogy to chapter 11 of Genesis, the individual pieces of the tower of knowledge would continue to be built but would no longer fit together. Another component of Bush's utopian vision for postwar American science was that continued disciplinary fragmentation was no longer universally considered the good it once was. As it turns out, Bush was right on the mark. For example, modern medicine, in which specialization reigns supreme, has made Herculean achievements in almost all areas of disease, surgery, and diagnosis. Yet such success, like all Faustian bargains, has not come without some cost. The whole patient has disappeared behind an array of charts, graphs, measurements, and specialized division of labor and materials.[88] Much like Henry Ford's assembly line, medicine stripped the patient (or the device) into components, took each systematically, and developed highly technical, efficient, and industrialized solutions that utterly ignored the whole organism.[89]

As the collapse of solely disciplinary identities has confronted scientists, so too have students of science studies encountered significant obstacles to understanding recent science because of the very complexity of the social bricolages in which scientists learn and work. The years since World War II have seen any number of scientists and engineers consciously working outside disciplinary boundaries. In a good example, Jamie Cohen-Cole noted that psychologists assembled an "interdisciplinary community of researchers" when gathering the resources needed to establish and promote Harvard's Center for Cognitive Studies.[90] Cohen-Cole illustrates nicely that much of the problem is attributable to the descriptive language used to cope with the unit of analysis, traditionally the discipline, institution, or the instrument. In recent science studies, descriptions of non-disciplinary forms of scientific practice meander through a variety of subordinate disciplinary labels, including 'discipline', 'branch', 'sub-field', 'interdisciplinary', 'multidisciplinary', and 'transdisciplinary'.[91] This indeterminate language has plagued historical understanding. It has also reflected the widespread problem of identity for Cold War–era scientists themselves. In hindsight, we recognize the convoluted identities of Cold War–era psychologists' genuinely evoking the experience of scientists forging complex new communities beyond the limited boundaries of disciplines.

Edge and Mulkay's study of British radio astronomy was one of the earliest examples of this struggle. They encountered terminological quicksand when attempting to denote what unit of science radio astronomy was. They called radio astronomy a branch, a specialty, a research community, a discipline, and a research network.[92] Their struggle to describe radio astronomy accurately was not so much a confusion of categories as a real recognition of the fuzzy nature of scientists, sociologists, and historians' accounts of the rapidly changing nature of the structure of science in the Cold War era. Science became difficult to describe precisely because scientists were transforming themselves and their social organizations into branches, specialties, communities, disciplines, and other identities. Likewise, increasingly fluid social and intellectual boundaries have spread far and wide. In the end, whether we want to understand the changes in multi-disciplinary science, in multi-national organizations, or in multi-conglomerate corporations since World War II, the litany of descriptors signals the immense challenges that face those trying to develop more community-based ways of thinking.[93]

A Single Sky explains how the open, international, and interdisciplinary social community of the radio astronomers created new knowledge. Community did not confine the radio astronomers to small-scale science, but permitted a new class of giant radio telescopes to be built outside the realm of the military-industrial complex. That such "big science" might emerge via cooperation rather than competition certainly struck radio astronomy's early supporters as unexpected, even supernatural. At Harvard in 1952, Harlow Shapley, the outgoing director of one of the United States' oldest observatories, said: "The rich harvest of exciting knowledge that radio astronomy has yielded in recent years has been gathered almost exclusively by electronic physicists, men skilled in phototubes, circuits, and the intricacies of electronic science. The discoveries made by these wizards of the micro-waves are largely in the field of astronomy."[94] Thus the old optical astronomer Shapley celebrated the notion that significant contributions could be made to a science by people without formal training, and using foreign equipment, via cooperation and the formation of a shared understanding. Recall the opening anecdote: The Welsh villagers could celebrate an English measurement of their mountain to declare their identity as Welshmen. A hill became a mountain through the cooperative transformative efforts of the cartographers and the villagers. Likewise, even though the radio wizards were disciplinary outsiders and their radio electronics magical, Shapley praised them and their wondrous discoveries because they had been made into "astronomers," and their instruments into "telescopes," via the making of an open cooperative scientific community.

In June of 1945 a powerful radio noise jammed all the receivers at New Zealand's Norfolk Island radar station. The operators feared a Japanese attack, but the latest receivers and direction finders identified the sun as the direction of the radio noise. Cooperation among the MIT Radiation Laboratory, Britain's Telecommunications Research Establishment, and Australia's Radiophysics Laboratory had raised speculation about the sun as a source of radio waves, and several outbursts had been detected during the war years. In Australia, the deputy chief of the Council for Scientific and Industrial Research's Radiophysics Laboratory, Joe Pawsey, had already performed some simple experiments involving horns and parabolic dishes pointed toward the sky through a window,[1] but this latest news of the sun as a source of radio noise came as a surprise. Providentially, the father of one of the Norfolk operators was Dr. E. Marsden, Director of Scientific Developments at New Zealand's Department of Scientific and Industrial Research. Between the wars, the elder Marsden had been involved in radio research on the ionosphere and even in some attempts to quantify the effects of lightning discharges on radio reception. He had met the new chief of Australia's Radiophysics Laboratory, Edward Bowen, at MIT's Radiation Laboratory, and now he casually relayed the New Zealand radar station's findings to his new antipodean colleague. Bowen was "quite mystified by the results because it appears that while thermal noise from the sun is expected at radio frequencies one would not expect to be able to detect it [at] 200 Mc/s."[2] Turning an antenna to the vacant sky was commonly part of determining the inherent noise level of the antenna itself. During World War II, receivers had become sensitive enough to be able regularly to detect noise from the sun or from space. As Bowen later recalled, "the matching of an antenna to free space became formalized in terms of the characteristic *impedance of free space* having the well-known value of 377 ohms."[3] That is, noise from space was acknowledged as background. This was precisely the kind of unanticipated outcome from wartime work that many leaders of science expected to provide a springboard for postwar research. Bowen encouraged his deputy, Joe Pawsey, to begin work immediately.

Joseph Lade Pawsey was one of the fathers of modern radio astronomy. Born in Country Victoria, he graduated from the University of Melbourne in 1929, then pursued a master's degree studying how lightning flashes and other atmospheric phenomena affected radio signals. At the time Australian universities offered no opportunities to undertake doctorates, so Pawsey went to Cambridge University's Cavendish Laboratory and worked under Jack Ratcliffe on the effects of the ionosphere on radio propagation, primarily on the variations in intensity and direction of waves reflected from atmospheric layers.[4] Pawsey discontinued his association with universities after 1934, and until 1939 he worked in Britain with EMI Laboratories on the development of television, particularly antennas.[5] At the outbreak of World War II, Pawsey was one of the first scientists to be offered a position in the new Australian Radiophysics Laboratory developing radar systems for the Australian military.[6]

In 1945, the New Zealand reports drew Pawsey's attention to a single body in the sky: the sun. Suspecting that the sun also generated most background noise, Pawsey sought to locate its exact source. On October 3, 1945, the Australian team of Joseph Pawsey, Ruby Payne-Scott, and Lindsey McCready first detected and localized solar radio noise.[7] Within three weeks they determined that the noise was correlated with especially heavy sunspot activity. The trio used a "sea-interferometer," a radar hut perched 400 feet above sea level on the South Head of Sydney harbor and armed with the latest radar receivers. The "sea-interferometer" combined two incident radio waves, the first directly incident on array of 40 half-wave dipoles and the second reflected off the ocean. The interference patterns formed by imposing the two signals on each other gave remarkably exact locations for the sources of the noise from sunspots as they traversed the face of the sun. In a paper published in *Nature* the following year, the Australians claimed that "the peaks of 1.5 metre radiation coincide with peaks of the sunspot area curve and with the passage of large sunspot groups across the meridian."[8] Though Pawsey sought to identify external sources of radio noise, which clearly had applications to the future reliability of radar, his project did not take that course. Once sunspots had been identified as radio sources, the work shifted to uncovering the mechanism of the production of radio waves from a star. This was the beginning of a new branch of astronomy, now known as radio astronomy.

If radar stations on Pacific islands seem an unlikely beginning for a story about astronomy, it may be because what we think of as "astronomy" has changed quite

FIGURE 1.1
The first Australian radio astronomy site, the Dover Heights radar station at Sydney, ca. 1947, California Institute of Technology Archives, OVRO 3.10–1. Reproduced courtesy of California Institute of Technology Archives.

radically since World War II. That change confronted the most famous astronomer of his generation, Edwin Hubble, who thought that, with the war over, he could simply "return to astronomy." During the war, Hubble had spent years away from the Mount Wilson Observatory and his beloved 100-inch telescope, then the world's largest. At the Aberdeen Proving Grounds in Maryland, he had led a team improving bazookas—a far cry from galactic astronomy, but illustrative of the great ruptures many scientists' careers experienced during the war. Hubble's prewar work established a fundamental relationship between Earth's distance from a galaxy and the galaxy's rate of recession away from Earth.[9] Hubble anticipated becoming the first postwar director of the expanded Combined Observatories, a powerful nexus of Californian people and instruments, including the 100-inch Mount Wilson and the soon-to-be-completed 200-inch telescope on Palomar Mountain, and also overseeing the expansion of astronomy at Caltech. The end of the war, however, disrupted Hubble's anticipation of a return to an idyllic astronomer's life. Hubble fretted that the Carnegie Institution of Washington, which oversaw Mount Wilson, might extend the wartime model of science and appoint as an administrator for the observatory someone with no feel for astronomy. His campaign against the administrators may not have been all altruism, but it certainly evoked an edenic vision of science.

Hubble feared "administrators." He told Vannevar Bush that oversight by an administrator who would run "the Observatory in a business-like way, decides (after taking advice) what problems should be attacked, [and] employs some astronomers to do the work" should be rejected utterly. At Harvard, the astronomer Harlow Shapley shared Hubble's fear that continued wartime arrangements would seep through American astronomy. Shapley ominously cautioned against his government's "intercession in American science": "Those who were worried [in the early twentieth century] about domination of freedom in American science by the great industries, can now [in 1946] worry about domination by the military."[10] Similarly, many of the young radio physicists at Australia's Radiophysics Laboratory, led by Ruby Payne-Scott, also railed against continuing classified work.[11] Not coincidentally, Caltech, Harvard, and the Australian Radiophysics Laboratory became major sites of the new radio astronomy.

As a war fought with gleaming B-29s, rockets, radar, and bazookas came to an end, Hubble romanticized the independence of the scientist at his instrument and alone with the universe. "Insofar as pure science is concerned," Hubble said, "leaders should be freed from as much as practical all affairs that do not directly pertain to research and research programs."[12] It is illustrative of how the world of science had already changed that Hubble's advocacy of purity struck the wrong chord with Vannevar Bush, both in his capacity as chairman of the Carnegie Institution of Washington and in his capacity as leader of the committee that would select the next director of the Combined Observatories. In October of 1945, the Carnegie Institution passed over Hubble in favor of the physicist Ira S. Bowen, a decision both Bush and Bowen knew would be controversial. They had agreed in advance on the content and the timing of the announcement. Unfortunately, word of Bowen's appointment leaked out prematurely and shamed Hubble publicly. As a concession, and since Hubble's studies of galactic structure were among the major reasons for the building of the Palomar telescope, Hubble would get a free hand scientifically and would have extensive access to the instrument. But Hubble remained embittered and isolated. Even his admirers noted that many people found his faux-English accent and his "boldness" grating.[13] When he wrote in a vicious letter that a physicist would "not be welcomed" as director of the "astronomical center of the world," Hubble's venom disturbed both Bush and Bowen.[14] Hubble could not have been more wrong, of course. Not only was a physicist welcomed, indeed embraced; the appointment of a physicist signaled a major shift in the culture of astronomy.

Hubble's failure to secure the Combined Observatories (which he called "astronomical center of the world") for his idea of an astronomer in the pursuit of pure science can be seen as a small example of the many changes that were already happening for science. "Science" as an endeavor was never the same after World War II. Those who tried to simply rekindle the old way of doing things were passed over, simply because many fresh young scientists were available. They, unlike Hubble, were not especially lured by the prospect of returning to old, run-down universities full of tired old equipment, professors, and ideas. Industrial firms in the victorious Allied countries sought talent far and wide and attracted large numbers of émigrés from traditional research institutions with equipment, personnel, and salaries that universities couldn't match. Prominent new national research agencies such as the Office of Naval Research in the United States, and established ones such as the Departments of Scientific and Industrial Research in Britain and Australia, recruited just as heavily. Universities, very much in third place, attempted to get back staff members and students they had lost to the war effort.

Before World War II, industry and national research establishments had been unattractive alternative paths for most scientists, being places of pragmatic technological development; after 1945, they commanded elite talent, the latest equipment, and budgets beyond the dreams of Solomon. Radio physicists were especially sought after. Their wartime experience with radar, with solid-state electronics, and with the production and detection of radio waves secured them especially appealing offers in the new field of avionics, and most went easily and lucratively to work for the military-industrial complexes in Britain and the United States. Most choose a path that, as David Kaiser has argued, similarly lured the nuclear physicists: the move to a culture of suburban science oriented toward service to the state.

The radio astronomers didn't emerge either from Edwin Hubble's dreams of scientific purity or from the easy allure of the military-industrial patron. They represent, rather, an underappreciated third way in the heady history of Cold War–era science. At first, the radio astronomers were little more than several isolated groups of radio physicists spread around the world and shying away from continuing to apply their skill and knowledge to merely the next radar system. This chapter emphasizes that those who would become radio astronomers were cut loose from highly focused wartime research agendas, then briefly wandered in the immediate postwar world in search of new directions. Collectively, the values they saw in one direction, and not another, would help characterize the radio astronomy community in the years to come. A scientific community dedicated to interdisciplinary cooperation and exchange, international cooperation and exchange, resistance to the military-industrial complex, and a belief that science can be the search for knowledge about the universe would forge the new radio physicists and the old astronomers into radio astronomers.

To understand the eventual culture of the radio astronomers' community, attention must be paid to the false starts, the novel job opportunities, and the ambitions of youth in these formative years. The radio physicists saw their wartime experiences as valuable and often romantic. The early optical astronomers interested in radio did not shy away from using the new technology to advance their science. Wartime exploits deeply affected how both radio physicists and astronomers viewed the pursuit of any science into the future. To understand the new science of radio astronomy and the new community of the radio astronomers, we must consider a strange discovery on a Pacific island, the appointment of a physicist to the "astronomical center of the world," and numerous apparently unconnected decisions by loosely affiliated people. Only by revealing this complex story of uncertain decisions can we explain the slow coalescing of a new ethos of science based on cooperation, openness, new technology, and interdisciplinary and internationalism—radio astronomy.

When we appreciate the choices various people made, we can understand that the loss of the directorship of Mount Wilson explains Edwin Hubble's venom toward Ira Bowen only in part. Rather, it was the loss of an entire culture of science that affected him so profoundly and made the appointment of a physicist so hard to bear. The second half of this chapter focuses on many of the issues that were at stake in Hubble's case. Significantly, a struggle for the support of "pure science" was repeated in the United States, in Australia, and in Britain. In each locale, as thousands of new scientists eagerly accepted new jobs in government labs and industry, some scientists saw their control of science slipping away and fought to protect it. All the while, government and military establishments—eager to recruit talent—argued that they wished only to support research, not to interfere. Under that presumption, several universities took radical new steps to include the new military-industrial complex, while others resisted any such collaboration. Decisions were just as problematic for institutions as they were for individuals during the crucial years 1945 and 1946.

Custer Baum was a young astronomer at Caltech at the end of World War II. One might think that astronomers are of almost no practical use, and that industry held little appeal for any astronomer. Not so in the Cold War world. As the imminent end of the war approached, Baum desired work outside astronomy, while industry lusted after him and those like him. If Baum had stayed at Caltech and worked under Ira Bowen at Mount Wilson, he surely would have had an ideal position in the world of science, especially with the 200-inch Palomar reflector nearing completion. Caltech had already begun planning the Jet Propulsion Laboratory, and California was awash in talent, energy, and opportunity. But, like many others, Baum left a research university and went to work for the Hughes Aircraft Company. He confessed to Ira Bowen in mid 1946 that he had also received an invitation from the Consolidated Vultee Aircraft Corporation to an all-expenses-paid interview and one from the University of California Radiation Laboratory asking him to work on cloud chambers, though the latter was "not of a permanent nature and not connected even remotely to astronomy from what I could ascertain." Nervous about his move to industry, Baum hoped "the actual job" wasn't too different what he had been led to expect. The bait was money. His recruiter, a Dr. Raymo, had told Baum to ask for "$6000 in the official application." No salary in astronomy was going to compete with that. (The director of Mount Wilson and Palomar was making only $10,000 a year.) "To my humble aspirations," Baum wrote to Bowen, "such a salary seems fantastic." The Hughes Corporation, Baum revealed to Bowen, hadn't even bothered to negotiate his salary claim.[15]

When we consider Baum's story alongside Hubble's, Pawsey's, or Payne-Scott's, we begin to see how, after ten years of depression and six years of war, many scientists embraced the notion of science as a job rather than a calling.[16] Hughes, Consolidated Vultee, the military, the government, and Berkeley all wanted scientific knowledge put to work for the security of the state, whether through aircraft, electronics, or atomic bombs. Edwin Hubble's vision for pure science, alone and uninterrupted on a mountaintop, contrasted with Custer's Baum's desire for financial security and willingness to enter corporate research. But the case also startlingly reveals much of the trouble with the emerging character of science. Baum was also offered a position in high-energy physics at Berkeley and some unknown job in industry. On what basis his expertise appealed to those two prospective employers we do not know. The offers made to him do illustrate, however, the links already forged between the military-industrial complex and high-energy physics. At the very least, those prospective employers wanted the same sort of expertise, and such offers took some of astronomy's devotees far away from astronomy. Even as Baum moved into industry, he recognized a distinction between astronomy and what is now known as the military-industrial-academic complex. Although Baum's departure and Hubble's demotion were evidence that old disciplines and institutions had lost their allure (and thus their power) in the face of new financial opportunities and new technologies, the idealism of pure science remained, even if only as a trope or a hope. From the perspective of an émigré at the point of embarkation, Baum saw the Hughes Corporation, the military, and Caltech all competing for the same people—people who, like him, seemed able to move from astronomy into something that, whatever it turned out to be, was definitely not astronomy.

From a single institution among hundreds, two men's choices spanned the postwar visions of science. Hubble looked to isolate astronomy from the administrators; Baum, now beholden to the administrators, feared that he would never encounter astronomy again. Baum hoped that his new job would be at least connected to the familiar world of the university; Hubble faced the prospect that a physicist would be put in charge of the world's foremost astronomical institution. Baum's experience and his choices remind us of the abundant opportunities that were available to anyone with scientific potential immediately after the war. That some people chose to stay at universities is even more significant in that light. The few who stayed, and the fewer who chose radio astronomy, ultimately changed the entire concept of man's place in the universe.

As an astronomer from Caltech found new employment as a physicist, a physicist became an astronomer. Ira Bowen, the new director of the Mount Wilson and Palomar observatories, opined that "new concepts or new experimental techniques in physics" heralded astronomy's future. Caltech possessed a far stronger ethos of disciplinary cooperation than most observatories. Its cooperative emphasis had been a central tenet of the institution since the days of George Ellery Hale, and partly explained why Bowen was chosen over Hubble as the first postwar director. Still, though Bowen had been trained as a physicist, he did not view the discipline of physics as coming to dominate astronomy. Bowen emphasized participation, openness, and cooperation among specialists. He envisioned getting the observatory's "teeth into fundamental new problems leading to new concepts." The "applications of nuclear physics to astronomical problems," especially the abundance of elements in stars, was the centerpiece of the research program that Bowen eagerly outlined for Vannevar Bush.[17] The glamour and allure of nuclear physics shone brightly after 1945.

The ideal of cooperation between subject areas was reinforced, at least in California, by the administrative merger of the Mount Wilson observatory (run by the Carnegie Institution) and the Palomar Mountain observatory (run by Caltech) into the Combined Observatories. Vannevar Bush termed the amalgamation "a new and important phase" for the observatories.[18] It was anticipated that "the research program of the observatories will be reinforced by studies on the campus of the California Institute, and graduate training leading to the doctorate will be given under the auspices of the California Institute by an astrophysics staff consisting of members of both the Institute and the Institution."[19] To what extent Caltech's model of cooperation and community would spread to become typical of astronomy as a whole remained a question, but the postwar influence of the Mount Wilson and Palomar observatories was considerable because of the powerful combination of superior telescopes and a new graduate school.

Examining Caltech's experience is a good way to begin to appreciate another important theme in the emergence of the radio astronomers: the process by which the Cold War mixed categories that only a generation earlier had been passionately delineated and defended—university and industry, pure and applied, physicist and astronomer. The best way to start is by overtly comparing Caltech against two examples in Great Britain, where, as much as in the United States, the military, industry, and the state scrambled to secure and utilize wartime research skills and novel technologies.[20] Many young British scientists, like their American cousins, sought financial security in industry, while others looked to pursue research less focused on immediate application. For instance, Jack Ratcliffe, a professor at the University of Cambridge's Cavendish Laboratory, wanted to talk to young Bernard Lovell about "Cosmic Ray bursts" in early June of 1945. Lovell had won renown during the war by developing the H_2S radar that was used to guide British bombers to their targets.[21] Looking to leave the wartime radar work behind him, Lovell embraced cosmic ray research, which sought to understand the properties of naturally occurring elementary particles and which stood as a viable alternative to the artificial production of particles that by then was taking place in cyclotrons. The stark reality of postwar Britain dawned on Lovell as soon as he re-entered the University of Manchester in 1945. During the war, Lovell had had a bomber available on standby for an experiment; now he found himself unable to obtain a new part to resurrect his old cloud chamber——a part that cost only a few pounds. "The glamour, the excitement, and the fear of [work after] 1935 [was] replaced by the loneliness and emptiness of the [postwar] laboratory," Lovell later remembered.[22] Both Lovell and Hubble desired to return to a scientist's life at the conclusion of their war work, but Lovell ended up lamenting what Hubble idealized: the scientist alone with nature and a broken instrument. Still, he chose to stay in his lonely laboratory.

After the war, the University of Manchester followed the rest of Britain into decline. For the most part, the University of Manchester built on its past substantial investment in the connection between science and local industries. Bernard Lovell's dream of returning to fundamental science, however, parallels the career of Alan Turing.[23] Lovell's memoirs notwithstanding, a tremendous anticipation had emerged during the war about the utilization of new technologies for future research. Super-sensitive radar receivers, improved aerial design, and solid-state electronics seemed very promising. The doyen of ionospheric physics, Jack Ratcliffe, had made Lovell aware of a small team working with the latest radar receivers. J. S. Hey's team had begun looking into some strange patterns of radar interference that had been irregularly observed from the sun by Britain's Army Ordinance Research Group. Ratcliffe thought some small cooperative endeavor between Lovell and Hey might be profit-able. Lovell, more interested in proposing rocket experiments for cosmic rays, con-tacted his former Manchester professor Patrick Blackett to resurrect an old idea they had had about radar echoes and cosmic ray showers.

Continuing wartime arrangements permitted some long-shelved experiments to reemerge, but the imminent dispersal of personnel and equipment required fast action. In order to learn more about cosmic rays (especially their interaction with the upper atmosphere) and to accurately determine the distance to the moon, Lovell wanted to borrow some 1.5-centimeter-wavelength receivers and some 13-centimeter-wavelength receivers from Britain's Telecommunications Research Establishment, and some people from the Anti-Aircraft Command, for six months.[24] But with the delays and shortages that the closing months of the war brought, Lovell was left calculating integrals in Manchester's old halls and trying to recall what his and Blackett's ideas had been. He despaired of the whole business of cosmic ray research without "close control of a station or better still our own apparatus with people to run it."[25] Still, once the difficult rocket telemetry calculations were put aside, Lovell and Hey did pursue the detection of radar echoes from cosmic ray showers for a short time. Man-chester itself then became an obstacle, as Lovell soon encountered radio interference from the activities of the city. Fortuitously, Lovell found a parcel of land well outside Manchester, at a site known as Jodrell Bank, that the university's Botany Department owned but wasn't using. Near the end of 1945, Lovell moved some radar equipment out to the radio-quiet site and began some observations.[26]

Lovell's interest in cosmic rays emerged as a consequence of his interest in physics and his expertise in radio receivers. Astronomy wasn't even a blip on his radar screen yet. Lovell valued fundamental research for different reasons than Vannevar Bush, who took a more pragmatic view related to industrial and economic development. As early as mid 1944 Lovell was of the opinion that, after five years of war, Britain had exhausted the potential of prewar fundamental research. Nearly an entire generation of people, he argued, had thought about research "deductively" rather than "inductively." Lovell attributed Britain's remarkable wartime scientific successes in radar, computing, and aircraft to men who had been trained in inductive thinking, but with the end of the war drawing near he argued that "the failure of education in the correct way of thinking and behaviour" was "the basic reason why we feel that our great days are over." In short, Lovell returned to the dim halls of Manchester to "restore the status quo of fundamental research" and do his "part in educating what little of the civilised world remains."[27] Pure or fundamental research was not merely the ambition of the scientist; it was a missionary calling to bring the light of civilization to the world.

In Cambridge, about 200 miles away, Malcolm Applebey, the director of research at Imperial Chemical Industries, invited another young radar researcher, Martin Ryle, to ICI's head office for the purpose of discussing "a suitable industrial post." The crystallographer Lawrence Bragg had forwarded Ryle's name to Applebey, but Ryle stalled ICI's offer just long enough so that Mark Oliphant, a nuclear physicist also casting around for bright young chaps to move to Birmingham to work on "fundamental nuclear physics," could all but offer Ryle an immediate position at least equal to an ICI fellowship.[28] Ryle's flexible postwar options parallels Custer Baum's experience in Southern California. Both were offered lucrative industrial offers and opportunities in nuclear physics. Ryle's father took the liberty of writing for advice to his son's wartime Telecommunications Research Establishment leader, A. P. Rowe. Evidently Ryle's service to Britain's war effort in commanding a group of scientists had been both significant and rewarding, but Ryle's father believed that his son had found himself at a "cross-roads" by early 1945. He believed his son should take two months off before beginning any new work, and thought he had earned the right to pursue some quiet research for a while.[29] The younger Ryle tentatively stepped toward ICI, requesting copies of his B.A. and M.A conferrals from Oxford. But by June he clearly had chosen not leave Cambridge. Still, to further illustrate the new mixing of categories in this fluid period, an ICI fellowship supported Ryle for the next three years at the Cavendish, permitting him both continued access to a research center and some financial security. Unlike Bernard Lovell at Manchester, however, Ryle received a sage piece of advice. Several years later, just after the death of his father, Ryle wrote a short note of thanks to Harry Plaskett, an old friend at the Oxford Observatory who had given him, he said, some solid guidance back when he was having to make difficult career choices. "I do not forget," he wrote, "that in the difficult early days after the war your advice helped me to make the decision to try and become an astronomer."[30] At exactly the moment when Custer Baum left astronomy, another young scientist had arrived at its doorstep.

In the United States, others were looking to move from amateur radio physics toward astronomy. Around the middle of 1945, Jesse Greenstein, a young astronomer at the Yerkes Observatory near Chicago, struck up an irregular correspondence with Grote Reber, a radio physicist at the Bell Telephone Laboratories in New Jersey. Reber looked to Greenstein to aid his plans for a giant radio antenna in the form of a dish to detect radio signals from stars and galaxies. In 1946 they collaborated on a paper.[31] Greenstein was the first professional astronomer to become interested in the curious radio noise that Karl Jansky had detected from the Milky Way region of the sky as early as 1937. In his youth, Greenstein had been curious about the antennas he had noticed at Holmdel during summers on the New Jersey shore. Greenstein (now a graduate student at the Harvard Observatory) and the Harvard astronomer Fred Whipple combined an interest in interstellar dust with the early radio surveys and concluded that large interstellar dust clouds were the source of the radio emissions.[32] Greenstein later claimed that Reber, an amateur astronomer with a homemade telescope, had "re-aroused" his interest in the problems of radio noise and directly "stimulate[d]" an influential *Nature* paper "about the free-free emission from interstellar gas."[33]

As Greenstein reached astronomy's professional summit by moving to Caltech in 1947, Reber's grandiose plans for a radio astronomy observatory complete with "engineer, astrophysicist, machinist, and laborer" amounted to naught.[34] The Harvard Observatory, among other institutions, turned down Reber's offer to build a $28,000 "radio telescope," even though the potential of radio techniques had attracted the attention of the galactic astronomers Harlow Shapley and Bart Bok.[35] Both Harvard and Caltech would build radio telescopes within ten years, but entirely without Reber. Reber's conception of radio astronomy and his idea of a radio telescope would be emulated, but after World War II astronomy didn't want amateurs. It wanted professionals, respectability, new telescopes, and disciples.

In the larger world of science after 1945, rocketry attracted serious attention, but nothing bedazzled quite like nuclear physics, that newly hegemonic Cold War science. For example, Dennis Robinson, once a lecturer at Birmingham, moved to the Electrical Engineering Department at MIT to work in a "venture to manufacture commercially 2-million volt Van de Graaff generators for cancer therapy, industrial radiography, and particle acceleration."[36] Bernard Lovell had hoped to bring Robinson to Manchester. Instead, Robinson was soon sending Lovell promising but surplus students. Likewise, Joan Freeman, one of the two female researchers at Australia's

Radiophysics Laboratory, left Australia to pursue nuclear work with Mark Oliphant in Birmingham in late 1945 (arriving, unfortunately, just before Oliphant left). She was eager to establish herself in an exciting emerging field.[37] In Britain, Patrick Blackett acknowledged that many other countries were interested in atomic energy mainly as a source of cheap power for industries[38]; India, now on the verge of independence, was a prominent example.[39]

Yet it is important to remember that, from the perspective of July 1945, radar had been the true wonder weapon of World War II. Only penicillin seemed its likely rival for acclaim in anticipated historical accounts. Unmentioned in Vannevar Bush's report *Science—The Endless Frontier*, one of the most significant documents of postwar planning, the atomic bomb project became merely an extra chapter tacked onto James Phinney Baxter's 1946 *Scientists Against Time*, the earliest historical account of the "wizard war."[40] For both Bush and Baxter, radar had set a standard for the measure and meaning of science for the postwar world. The conduct of science, its patronage, its recruitment and training, and its place in the postwar economy would be modeled on the experience of radio and radar physicists in the United States at the MIT Radiation Laboratory, in Australia at the Radiophysics Laboratory, and in Britain at the Telecommunications Research Establishment.

New research programs grew easily and rapidly for the radio physicists. In 1945, in California, Luis Alvarez began to use old radar sets in series to construct a linear accelerator. In Australia, the Radiophysics Laboratory's new chief, Edward Bowen, had the much the same idea. Both Alvarez and Bowen had been foundational members of the massive American radar research effort, the famous "Rad Lab" at MIT, after 1940. Subsequently throughout the war, the American Rad Lab, the Australian Radiophysics Laboratory, and the British Telecommunications Research Establishment were connected through a series of Scientific Liaison Offices in London, Washington, and Sydney. Complete copies of all radar-related reports available in all three locations bolstered regular personal exchanges. In contrast to the secretive, compartmentalized, barb-wired Manhattan Project, and later the National Laboratory system of the American nuclear program, radar's researchers inventive ethos emerged from the war comfortable and experienced with international cooperation and exchange, and beholden to the idea of pure research as precursor to successful applications.

Cooperation and international exchange shaped the creation of a powerful but cheap electron accelerator by the Australian Radiophysics Laboratory in Sydney, for example. Early in 1945, that lab's leader, Edward Bowen, assigned one of his most promising young researchers, J. S. Gooden, to "design a low voltage model to test the principles involved" in electron linear acceleration.[41] Gooden came from the Melbourne University section of the Radio Research Laboratory, a prewar group devoted to the study of atmospherics that lost most of its staff to the CSIRO's new Division of Radiophysics in Sydney in the war years. Gooden completed his model by April. The results "looked so hopeful" that Bowen had him "start right in on the construction."[42] Acceleration of elementary particles was a crucial opening step in nuclear research, which Australia already valued for her future national development. Independent research was necessary, however, because as early as February of 1945 the British chancellor had told Mark Oliphant that Australia would not be given "information of any kind on this particular matter."[43]

Bowen and Gooden's project was a much smaller version of a similar project in Berkeley. Luis Alvarez had amassed nearly 3,000 SCR-268 anti-aircraft directional radars that the US Army had discarded after the introduction of the new model.[44] While he made technical progress, Alvarez's plans for a "linac" (linear accelerator) remained largely "frustrated" throughout 1945 by his opposition to the Army's wish to fund the project through the Signal Corps, which would entail leaving Berkeley for New Jersey.[45] Enmeshed in international networks, Bowen soon became aware of the limitations of his accelerator project; soon afterwards he halted most of the work. Moreover, Bowen and Alvarez both shied away from continued military cooperation in early 1945 and preferred to search for fruitful avenues of "pure" research that might emerge from wartime technical advances. We know that the US military, most prominently through the Office of Naval Research, had already begun canvassing universities and institutions across the US for scientific talent. The effect of such patronage on science was immense. Scientific talent and resources were funneled into larger centralized projects, and smaller efforts were abandoned. But scientists gained immense esteem as they worked to build the next generation of wonder weapons and to defend democracy against fascism and later against communism.

Bernard Lovell, Martin Ryle, and Edward Bowen all emerged from Britain's experience of chaotic panic in 1940 and 1941 to lead various sections of that country's wartime radar effort. By the 1960s, these three men would be the leaders of three of the five most significant radio astronomy projects in the world. We might expect that such men, who had seen victory emerge from their efforts, would have embraced or even openly advocated the closer military-academic-industrial connections. Instead, they chose to embrace astronomy as a career and an identity—astronomy, whose own practitioners had often found it difficult to be of use during the war years.[46] Their experience of World War II was one in which one "diversion" after another distracted researchers and constant movement disrupted any actual development or installation work. Romantically, they became known as "boffins." A boffin's "lab coat was replaced by the bulky flying suit [and] a parachute which also served as a seat."[47] They fitted Blenheims one day, French aircraft the next, and Beaufighters the next. With no training manuals, they and others like them employed makeshift techniques and worked under intense pressure to introduce complex scientific equipment into a military establishment.[48] The three were but a small part of the vast assemblages of international and transdisciplinary practitioners brought together by the contingencies of an Allied war effort that spanned the globe. It was radar that had unified the practice of Alvarez, Bowen, Lovell, Blackett, Ryle, Bush, and Ratcliffe through the years and laid the foundations for the community-style science of future radio astronomers. Bowen and Lovell would join Ryle as astronomers, but Ryle's fraught transition from a radar physicist working at the Cavendish Laboratory supported by ICI grants to a Nobel-laureate astronomer exemplifies deeper changes that affected the post-1950s generation of scientists. Ryle, like the chemist Linus Pauling, would become highly skeptical toward and even openly critical of the close relations between the military-industrial complex and nuclear physicists.

To coherently assemble the fragments I have been outlining, and to grasp the shape of the radio astronomy community that would grow after the close of World War II, one might take an epidemiological approach. Essentially, Luis Alvarez was radio physicist zero. Alvarez—a prominent member of the MIT Radiation Laboratory until 1943, when he left to work on the Manhattan Project—reappeared to build linear accelerators after 1945. Alvarez worked closely with Edward Bowen; indeed, he would later describe Bowen in his memoirs as "the most important person to join the [Radiation Laboratory]."[49] Bowen, one of the earliest members of Robert Watson-Watt's small team working on radar in the late 1930s, possessed a complete working knowledge of all the wartime British developments in radar, especially airborne intercept radar and the cavity magnetron. Bowen was a member of the select team that accompanied Henry Tizard to the United States in 1940 to deliver every important technological device in the British arsenal into the hands of American industry. In fact, Bowen carried the suitcase containing the prototype magnetron across the Atlantic. However, Bowen had volunteered to go with Tizard partly because A. P. Rowe had been elevated above him within Britain's wartime radar program.[50] Rowe thus became Martin Ryle's group leader in the latter part of the war, while Ryle and Lovell were brought into contact through Jack Ratcliffe and Patrick Blackett, their respective heads at the Cavendish and Manchester, by questions concerning the use of rockets in cosmic ray research. Having worked in the United States between 1940 and 1943 before moving to Australia to head that country's Radiophysics Laboratory in 1945, Edward Bowen met, and deeply impressed, Vannevar Bush. Bowen's correspondence with Bush always maintained an air of writing to an old schoolmaster, while Bush's replies were overtly friendly, open, and personal, always beginning "Dear Taffy." Bush was, among his many significant roles, chairman of the Carnegie Institution; thus, he was Ira Bowen's boss, and Bowen was Edwin Hubble's boss. Ira Bowen was the confidant whom Custer Baum told of his impending departure from Caltech; he also supervised the appointment of Jesse Greenstein to head the new graduate program at Caltech. Greenstein was a sometime correspondent of Grote Reber, and was a graduate of the Harvard Observatory. Greenstein and Leo Goldberg had mentored by Bark Bok and Donald Menzel, who would shape Harvard's radio astronomy program after 1952. Though these men drifted to California, Australia, Cambridge, and even Manchester by 1945, the strands of private correspondence, mentorship, friendship, and even animosity helped them to build a new scientific community.

Those small streams of individuals unsurely groping their way forward into peacetime joined broader rivers that would deeply affect all sciences after 1945. Patterns established during the "wizard war" brought long-lasting and deep changes to the character and conduct of science including the embryonic radio astronomy community. The first program for change, and certainly the most famous, was sent to President Harry Truman three weeks before the Trinity test: Vannevar Bush's report *Science—The Endless Frontier*, which called for huge increases in America's support for basic research (Bush's term for pure science).[51] Science, technology, industry, and the academy, not to mention American military might, had flourished with state intervention and support during World War II. America's wartime strength, Bush argued, had come from taking established scientific ideas and creating benefits for its population. "Most research in industry and in Government," Bush wrote, "involves application of existing scientific knowledge to practical problems." And so Bush issued an important caveat to any hope for continued economic and technological prosperity: the only way to continue the economic, social, and defense benefits of science into the future was for the US government to promote a pattern of supporting undirected research. Previously, the US had relied on, and had benefited greatly from, pure research—almost all of it conducted elsewhere, especially in Europe. It now fell to the United States to "increase the flow of new scientific knowledge through support of basic research, and to aid in the development of scientific talent."[52] In short, the US needed new researchers in every conceivable area in order to promote new industries, new jobs, and new knowledge for defense.

Science—The Endless Frontier is a snapshot of the achievements of science in wartime, with the atomic bomb left out. Penicillin and radar were the wonder weapons of science that had made victory possible. Both had come, Bush argued, not from prioritized research, but from independent study. One of the more powerful lessons derived from the war was that seemingly esoteric investigations had, often surprisingly quickly, transformed the battlefield. Penicillin and the cavity magnetron were icons of this argument. Over and over again one encounters the story of Mark Oliphant at Birmingham and the magnetron (a cavity resonator able to deliver powerful pulses of microwaves for radar). Bush viewed basic research as the foundation upon which all other products of science, especially technological ones, were built. Bush had to refine his idea of the place of "basic research" within science and the process whereby it would produce useful technologies or industries. In a series of talks, appearances, and public lectures, Bush thought it necessary "to explore this word research":

Basic research in any field, and especially in natural science, is severally described as the exploration of the unknown, as the search for new knowledge, as the investigation of problems whose existence but not whose nature is realized. In terms of people—and these are by far the most important terms—basic research is the focusing of a keen trained intelligence by a patient and unsparing will on the problem of discerning the unknown, then bit by bit defining it, and finally bringing it into the range of the cognizable. Once basic research has achieved its purpose, and has translated an unknown into the known, applied research may be called on to reduce the new knowledge to practice, to make it available for man's direct use as part of the physical, intellectual, or industrial apparatus with which he controls his environment and better his life. . . . He must be able, and he must be free, free to put his ability to use as only he can do.[53]

Here was a powerful ideological vision of scientists freed from military or industrial obligations and unbounded by notions of utility or usefulness. For the pragmatic United States, not to mention Australia, it was little short of anathema. Yet the complete displacement of European scientific talent immediately before and especially during World War II impaired American access to new knowledge, Bush argued. Without that large reservoir, American applied science must establish its own sources of basic science. Bush thus envisioned a large established federal agency to distribute money wherever scientists saw fit. We know now that Senator Harley Kilgore, a Democrat from West Virginia, attacked and defeated Bush's vision. Granting scientists unlimited budgets in combination with autonomy seemed to trade away federal control over the new agencies, and Bush's argument about the conversion of "basic science" into applications swayed too few members of a Congress that was looking to regain fiscal control.[54]

Among the practical consequences of pragmatism were the emergence of the military-industrial complex and the general reshaping of American science. The US Army Signal Corps, for example, supported both MIT's Research Laboratory of Electronics and the Columbia Radiation Laboratory. The Columbia Radiation Laboratory received nearly a quarter of a million dollars a year from 1945 on, using it to support six faculty members, a technical staff of twenty, and about twenty graduate students.[55] The Signal Corps was investing not only in available manpower but also in a sustained manpower program designed for the unforeseeable future. Because the military envisioned an extremely long, protracted new "wizard war" with the Soviet Union, temporary laboratories and crash programs would not suffice. Only permanent, self-regenerating, military-sponsored laboratories would fulfill the needs of the warfare state.

The examples cited above hint that the creation of new scientific talent was one aspect of Vannevar Bush's vision that survived.[56] Bush suggested that "unless the ranks of scientists were replenished and the funds for basic experimentation were increased, American innovation might cease."[57] According to *Science—The Endless Frontier*, scientific knowledge now depended on large-scale recruitment and training of new scientists and engineers. To continuously renew science so as to provide for prosperity and security, Bush advocated support for the major centers of basic research— "principally the colleges, universities, and research institutes"—where "stability of funds" for "long-range" projects were enshrined.[58] Older institutions, such as Harvard, Yale, and Princeton, did little to encourage the postwar science-funding boom, often entering into patronage relations with the federal government only with great reluctance. However, they acted to ensure openness as opposed to secrecy, especially in the production of graduate theses.[59] Lee DuBridge's candidacy for the presidency of Caltech in 1945 rested, his supporters claimed, on his unwavering commitment to the full independence of academic institutions in the face of bountiful government aid.[60] American academic institutions steadily adapted their undergraduate and staff recruitment and modified the training of graduate students to serve and reinforce ties with the military-industrial-academic complex in a feedback cycle.[61] Stanford and MIT took early advantage of military support for their expansions, while Johns Hopkins established a separate research laboratory, the Applied Physics Laboratory, well away from any established instructional functions.[62] This emphasis on the importance of expanding the base of scientific talent, rather than any direct research agenda, shaped the radio astronomers as much as anyone else.

The physicists have long been seen as the primary beneficiaries of the military-industrial-academic complex, with ballooning numbers, laboratories, and budgets. In the face of abundance, most physicists gorged on the military-industrial complex, leaving many astronomers looking pale and undernourished indeed. The story of the physicists' prosperity remained occasionally tempered by deep concerns over the loss of their cloistered prewar world. Rare voices mourned the "sad result of the relentless pressure of numbers," as Raymond Birge openly wrote.[63] Daniel Kevles could celebrate the physicists' "victory for elitism," but the post-Vietnam generation drew far different conclusions about how prosperity had affected physics and physicists. The physicists seemed too eager, indeed gluttonous, in their embrace of federal patronage, and later recognized the dangers of trading autonomy for big budgets and bigger accelerators. After the excess of the postwar generation, the next generation experienced a famine. By the early 1970s, more than 1,000 physicists fought over only 53 jobs in the United States.[64]

Into the 1960s, the powerful union of radio physics and astronomy that created "radio astronomy" appealed to the two central contexts of the Cold War era: pure science and the recruitment of disciples. In contrast to the physicists, the radio astronomers, like the optical astronomers, saw themselves as going comparatively hungry. Astronomy programs did not attract new cadres of students, nor did they receive vast research and development budgets. Although radio astronomers held to Vannevar Bush's idea of "basic research" as a secure ideological foundation, the new science still had to attract students. The creation of lures for students shaped the community-style science of radio astronomy. In short, if Cold War physicists built steeples, the radio astronomers established congregations.

The Commonwealth allies of the United States embraced Vannevar Bush's argument about the need to invest in pure research in order to ensure economic and industrial progress in a way that the Americans did not. In both Australia and Great Britain, community-style sciences such as radio astronomy used Bush's program to legitimize support for sciences without practical objectives, as well as appeal to new disciples via the argument that science should be interdisciplinary as well as international.

Many aspects of Bush's program took root in Britain and Australia far more readily than in the United States because Bush's vision resonated with local ideals of "science for its own sake." Plans to reassess the place of science existed everywhere, notably in Australia, Britain, and Canada. Australia's Labor government began planning for peacetime as soon as the Japanese advance had been halted, in mid 1942, inaugurating the Department of Postwar Reconstruction to oversee the transition to peacetime.

That department consisted of four major planning agencies—the Commonwealth Reconstruction Training Scheme, the Commonwealth Housing Commission, the Postwar Rural Commission, and the Secondary Industries Training Scheme—that gave priority to training and/or retraining in the envisioned future.[65] In Australia, as in all industrial nations, the Great Depression loomed over postwar planning; thus, the goals of postwar reconstruction were "to control the fluctuations of the market economy, maintain employment . . . and protect the unfortunate by creating an adequate system of social security."[66] Herbert Coombs, Australia's Minister for Postwar Reconstruction, sought to maintain the relative industrial harmony of wartime by appealing to each worker as a "citizen in a productive society."[67] Like Britain under Clement Atlee, postwar Australian governments concentrated on economic security and stability but also implemented many long-delayed social reforms.[68]

Postwar science often shared its governance with models of economic and industrial progress. The United States dispersed federal scientific demand and talent into established universities and industries and promoted entire new institutional behemoths to pursue research and development, thus concentrating the financial and technical power Congress had attempted to disperse by defeating Vannevar Bush's plan for a National Science Foundation immediately after World War II. Britain and Australia, on the other hand, concentrated their wartime scientific efforts into established and expanded government laboratories. In Britain, much of the work was done through the Department of Scientific and Industrial Research, the Australian counterpart of which was the Council for Scientific and Industrial Research. The British DSIR and the Australian CSIR openly pursued Bush's notion of science's driving future industrial and economic prosperity. In Australia in the 1930s, according to C. Boris Schedvin, "many politicians would have regarded CSIR as marginal to the economic welfare; a decade later it was seen as an indispensable national institution."[69] After the war, economic stability and security were central to the CSIR's mission. Postwar planning expanded science's anticipated role in Australian industry. John Dedman, Minister for CSIR, saw the organization's role in the postwar world as substantially different from its original charter as a solely "scientific" institution. "The war itself has convincingly demonstrated," Dedman said, "that the only way in which [industrial] adaptability can be achieved is through the proper application of scientific knowledge. [The CSIR] must be more than a great scientific institution—it must [bring] together the official or industrialist who has a problem to be solved [and] the scientists who can solve it."[70]

The cooperative wartime successes of Australian scientists working in concert with the British and the Americans without being subservient to them deeply impressed the highest levels of the Australian government. As in the United States, the formula that had worked so well during wartime saw scientists on call to answer the problems of business, government, or industry. After the war, the chiefs of the Australian CSIR supported scientists pursuing basic research that might, potentially, deliver new knowledge to industry, where it could be developed into applications. While Minister Dedman maintained that the CSIR was an "integral unit of industrial society,"[71] the CSIR's chairman, David Rivett, steadfastly advocated Australian scientists' "first-class capacity for original, deeply fundamental work." "It is a duty," he said, "to ourselves and to the world to give them every opportunity in their own land to contribute their best."[72]

As Australia's major scientific outlet, the CSIR thus became embroiled in the same debate that Vannevar Bush had lost: the debate about how basic research would do more than applied research to ensure prosperity. Immediately after the war, the physicists' vision of basic research undergirded the establishment of the Australian National University as a place "more attuned to managing domestic and international uncertainty" and envisaged "to complement the programmes of the CSIR."[73] Mark Oliphant and a costly cyclotron (later aptly named the White Oliphant) dominated the Australian National University's School for Physical Sciences. Like its counterparts everywhere, the White Oliphant exemplified Bush's warning, expressed as early as November of 1945, that nuclear physics might ride roughshod over astronomy, or indeed over all other sciences. "There is no question whatever that there will be enormous emphasis in [the United States] on nuclear physics, perhaps too much emphasis, and that this will reflect on the point of view of astronomers," Bush presciently wrote to the new director of "the world's center for astronomy," Ira Bowen.[74]

Much of the distinction between the American and the British and Australian attitudes toward support for basic research rests on acknowledging that the Australian and British postwar governments embraced the conception of investment in basic research not only as a marker of status but also as the fulfillment of the promises of a generation of wartime work. In Britain, Clement Atlee's Labour government stormed to power in the general election on the promise of a socially secure state system of employment, universal health care, and housing.[75] Like Atlee, Australia's new postwar prime minister, Ben Chifley, committed the nation to addressing outstanding social problems as the reward for six years of war. British and Australian scientists anticipated basic research opportunities at the conclusion of hostilities as part of that social contract. In contrast, the rapid shift toward military funding for science funded by the military-industrial complex in the United States, especially in the physical sciences, asymmetrically skewed American scientists' expectations, and the defeat of Vannevar

Bush's program buried it. Both the British DSIR and the Australian CSIR, however, moved steadily toward increasing the amount of basic research supported by their institutions, which were, in turn, supported by their governments. In contrast to the United States, as the CSIR's chairman, David Rivett, acknowledged, "CSIR would have to carry considerable and increasing responsibility in connection with fundamental research in Australia" because of the perceived weakness of the universities.[76] That responsibility paved the way for the support of radio astronomy in Australia.

Rivett's immediate subordinate—his chief executive officer, Frederick W. G. White—drew an even closer parallel between postwar Australian scientific ambitions and *Science—The Endless Frontier*. In one of the major planning statements outlining postwar Australian science, White declared the CSIR "responsible for initiating scientific work to ensure that Australia contributes to and can assimilate fundamental scientific knowledge from any source overseas." In what sounds almost like an echo of Bush, though it was published a year earlier than Bush's report, White asserted that Australia had "relied too long upon fundamental scientific work carried out overseas."[77] Frederick White, a New Zealander by birth, had done his Ph.D. work at Cambridge on upper-atmosphere radio waves, and had become the chief of the newest division of the CSIR, the Radiophysics Laboratory, in 1942. When promoted to the position of CEO of the CSIR, in 1945, he had been succeeded by Edward Bowen.[78] Mirroring Mark Oliphant's work on the cavity magnetron and radar in Birmingham, the work of Australia's own Radiophysics Laboratory, White argued, had aided the efforts of local science and industry in the dark days of 1942, supporting the manufacture of munitions and aircraft alongside steelmaking, shipbuilding, and chemical production.[79] And now, looking to the future, White duplicated the argument Vannevar Bush would make to President Truman, even including the corollary that new sources of fundamental knowledge would have to be created locally. "It is obvious," he concluded, "that applied science itself can prosper only if the fundamental aspects of the science are sufficiently appreciated and investigated. Any laboratory within CSIR must therefore devote part of its efforts to fundamental investigations and part to applied work."[80]

The beginnings of radio astronomy were far from salubrious, and certainly no one anticipated that giant radio telescopes costing millions of dollars would appear only a decade later. Only with hindsight, that powerful ally of history, can we recognize the early members of the radio astronomy community: Joe Pawsey sticking his head out of a window in Sydney, Vannevar Bush arguing for pure science, Bernard Lovell in the middle of a muddy field near Manchester, "Taffy" Bowen and his accelerator,

Jesse Greenstein talking with odd radio enthusiasts, Harlow Shapley taking a stand against the military-industrial complex, Martin Ryle laboring for ICI at Cambridge. None could, as yet, give the subject of their research a name (as we shall see in the next chapter, it was called simply "noise" for a time), nor did they know where it would lead in the years ahead. Yet the beginnings of a field in which astronomical knowledge would be gathered by means of radio antennas were forged in the immediate postwar years chiefly by this international group of men joined by wartime experience and brought into contact via other important characters in the history of Cold War science, especially Vannevar Bush. Bush was not making the decisions about the radio astronomy's new discoveries, or about instruments, but his central role as a visionary proponent of pure research and his critical social role as a node in the emerging network have been noted time and again. His schema for science's future places him as foundational for the institutional and pedagogical structure of the radio astronomy community. The radio astronomers took on many of the organizational and social aspects of Bush's vision for science, which were more widely embraced in Australia and Britain than in the United States. Following Bush, an emphasis on basic research and on recruitment and training of scientific talent became core features of the new community even as early as 1946, germinating in the confused and uncertain transition from wartime to peace. British and Australian radio physicists turned to any project that promised to validate their wartime experience. Accelerators and radio astronomy thus emerged as two sides of the new relationship between science and the state in the Cold War. While accelerators easily and obviously promised advances in nuclear technology, radio astronomy took advantage of the moral uncertainty of supporting only military or industrial applications of science. In Australia, in Britain, and in the United States, insecurity about the place and the importance of pure science permitted a small coterie to take highly practical radar devices and point them randomly at the sky in search of new knowledge. Edwin Hubble, who died in 1953, missed one of the most momentous changes that astronomy would undergo since the development of the optical telescope.

The Yanks are spending vast sums (unit = megabuck = 10^6).

—Frederick White, 1946

"We must provide for research in atomic energy," the retired Chairman of Australia's Council for Scientific and Industrial Research, Sir George Julius, was quoted as having said by the *Sydney Morning Herald* in late 1945.[1] After the successful use of the atomic bomb at Hiroshima, independent nuclear programs rapidly sprang up in Britain, Australia, Canada, Switzerland, and, of course, the Soviet Union. At the CSIR's Radiophysics Laboratory in Sydney, Edward Bowen anticipated an aggressive national effort and offered his entire linear accelerator group to "make an exceedingly strong team [which], given suitable assistance, would probably do all that is required to start Nuclear research in Australia."[2] A copy of the Smyth Report arrived in mid August, and Joe Pawsey, Bowen's deputy, outlined the American triumph to the rest of the Australian lab soon afterward.[3] Pawsey (who had been a graduate student at the Cavendish Laboratory in the heady days of the 1930s, when Ernest Rutherford, Mark Oliphant, James Chadwick, and Peter Kapitza all strove to break open and understand the nucleus) was as qualified as any outsider to the Manhattan Project could have been to disseminate the principles involved. But no great call rallying Australian scientists to develop fission was sounded. Instead, complete American secrecy surrounding atomic research stalled Australian (not to mention British) dreams of local atomic industries. The members of the Grand Alliance that had won World War II learned that they were not really equal partners but only tolerated members of the United States' nuclear club.

Bowen's linear electron accelerator remained the sole Australian development in particle acceleration by the end of 1945. Design and testing of the accelerator had been under way since March. By September, only a month after the war ended, the equipment had produced energies in excess of a million electron volts, considered the upper limit of the theoretical predictions for the single-cavity device. Bowen, directing the program, increased the number of people attached to the accelerator and promptly advertised Australia's success in the journal *Nature*: "[Our] results are sufficiently encouraging for us to proceed with the acceleration of electrons to considerably higher energy-levels by multiple acceleration."[4] Optimism ran high for the first year of the Australian project. Bowen's report to the government for the year to June 1946 gave "indications that this [research] is likely to produce a device of performance comparable to a moderately sized cyclotron but of simpler and cheaper construction." "The importance of such a machine," the report continued, "cannot be over-emphasised."[5]

Secrecy shrouded the vast extent of the American efforts from the Australians. Luis Alvarez's linear accelerator at Berkeley, for example, started with the transfer of $250,000 worth of surplus radar equipment, then received a donation from General Leslie Groves and the Manhattan Engineering District. "We ran it with a big barrel of greenbacks," Alvarez later said.[6] Alvarez's Californian competitor, the Hansen-Ginzton group at Stanford, started their work in 1946 with an initial budget of $65,000, successfully developing more powerful klystrons from discarded radar equipment.[7] And linear electron acceleration was the poor cousin to Ernest Lawrence's Radiation Laboratory at Berkeley. Lawrence accepted a $7 million budget from the Manhattan Engineering District *for 1946 alone* to support a staff of more than 200, including 66 scientists.[8] By comparison, at its height the Australian accelerator had a staff of nine and a budget of less than 20,000 Australian pounds—at least two orders of magnitude less than Alvarez and Lawrence. The historian Paul Forman has correctly called attention to the riches showered on American physicists by the Office of Naval Research, the Atomic Energy Commission, the Signal Corps, and other military funding organizations. The organizations of the military-industrial complex radically altered the agenda of vast numbers of working physical scientists, and the outstanding question has remained "To what end?"

The Australians informally learned of the extent of American efforts in linear electron acceleration in June of 1946, as Bowen's optimistic predictions went to print. The CSIR's executive officer, Frederick White, cabled Bowen an early warning that "nuclear energy work" was "all in a tangle." White advised that Australia "certainly should not set up any expensive gear for accelerating particles. The Yanks are spending vast sums (unit = megabuck = 10^6) and seem to be finding that linear accelerators are best and cheapest." In the face of overwhelming American dollars, White supported Bowen's decision to abandon the linear accelerator project. "It is very doubtful whether you will get very far in competition with the work overseas," White lamented.[9] And so the Australian electron accelerator all but vanished within two years, a small incident in the history of science in a faraway country.[10] Vannevar Bush, doyen of American science policy, dreamed of fluid movement of knowledge from science to industry, and Bush's model provided Bowen with a simple way to dispose of the entire business: the Australian government announced that "the [accelerator] work has progressed to a stage at which it can be taken over by a manufacturer."[11] This wasn't the linear model in operation; it was a convenient way to rid Bowen's Radiophysics Laboratory of a boondoggle.

The episode proved crucial to the emergence of radio astronomy as a science. Though the Australian efforts in particle acceleration might seem an obscure historical moment, the more generic aspects of the story sound familiar themes having to do with Cold War–era science: the excitement surrounding nuclear physics work, the money involved, the secrecy. We also begin to see that from 1945 onward the massive outlays of money both promoted new projects and served to eliminate smaller players, especially those outside the United States. Since the military-industrial complex possessed most of the available funding, it inevitably shaped much of science's direction. However, it is worth emphasizing that the military-industrial complex was not the only game in town (there were sources of funding outside the complex), and that the complex required working relationships with scientists and thus had to accommodate certain social norms (especially expectations of publications and secrecy). Quite inadvertently, the complex's rapid dominance actually encouraged research in areas that were outside its purview. When Bowen decided to abandon the linear accelerator, for example, he could direct the considerable talents of his laboratory toward alternative paths. One path that seemed particularly promising was the study of radio noise from the sun and stars, otherwise known as radio astronomy.

In 1946 and 1947, many smaller projects, like Australia's linear accelerator, were consigned to the dustbin of history. In those years, the leaders of the future radio astronomy community confronted a cacophony of competing agendas and visions for science. Astronomers complained loudly about the increasingly limited opportunities for new optical telescopes while insistently calling for rockets, radio receivers, and other new devices. A constant din came from old wartime social and professional networks. Friends, colleagues, and competitors now spread around the Western world offered advice and guidance. And all the while a nearly deafening roar howled from national and military institutions. Noise, in the radio sense, is random; in this crucial period, the astronomers and the radio physicists had to order and make sense of the random noise of agendas, patrons, and the heavens. But before the radio astronomers built a community that traversed disciplinary and national boundaries, they nurtured local radio and astronomical expertise, and actively listened and made sense of the sounds of the other. In making themselves both heard and understood, the radio physicists and the optical astronomers envisioned new astronomical instruments and invented strategies for recruiting new disciples. Local groups in various places joined together, reenacting the same process of listening and understanding to create a global interdisciplinary scientific community.

The Americans may have dominated nuclear physics research, but they couldn't work inclusively with an international community, and in effect every nation worked in isolation to develop an atomic bomb. Lavishly funded but altogether secret, atomic labor narrowed the scope of many scientists' vision of the horizon of knowledge. In distinct contrast, the formation of the radio astronomy community required the cooperation not only of individuals but also of disparate groups in Australia, Britain, the Netherlands, and the United States. The radio astronomers emerged when several groups of scientists coalesced around a set of new research tools—highly sensitive radio receivers—which became understood as new astronomical instruments. As we saw with the Australian linear accelerator, old radar sets could be reconfigured into particle accelerators. But the technology of radar could also be changed into the instrumentation of astronomy. The central question became not the form of the technology but the social shape of the people who would use it. The social shape of radio astronomy was an open, international, and interdisciplinary community. This chapter begins to explain how that shape came about. In short, it went as follows: Traditional optical astronomers already formed a tight, intimate community. The new radio physicists' associations were more recent, but also more sharply tempered by through wartime labor. Scientists of many stripes and disciplines looked forward to the potential of new instruments and new patronage to extend old research programs or to uncover new discoveries. Thus, new research programs embraced the cooperative ethos of the war years along with the easy flexibility of rapid succession through related problems.

Even as American cash swamped "Taffy" Bowen's dream of accelerating particles at Australia's Radiophysics Laboratory, Bernard Lovell moved his radar trucks into the middle of a muddy field at Jodrell Bank outside Manchester and tried to detect cosmic ray showers. Neither Bowen's nor Lovell's research efforts concentrated on anything that could be termed radio astronomy in the first few years after the war. Though we now know that Bowen's ambition to possess a major accelerator group could not compete with American megabucks, and that Lovell's cosmic ray ambitions were simply flawed, at the time they appeared to be profitable lines of research. They utilized both men's wartime expertise, and they were of immediate interest in the wake of the nuclear research that had produced the atomic bomb. It was as much, if not more, the failure of Bowen's accelerator and of Lovell's cosmic ray program that set the scene for both radio astronomy efforts.

Lovell used the same technology as Bowen, as did Alvarez. All three took advantage of piles of leftover wartime radar sets, quickly outdated by newer models, some not even out of their wrappings before being discarded—quite a metaphor for the next half-century of American materialism. Lovell, Bowen, and Alvarez possessed experience in antenna design, circuit noise, and early solid-state electronics; they looked at radio noise as displayed on screens or as printed out on chart recorders.

Lovell first began recording sightings of what he believed were reflections from the ionized trails of cosmic ray showers toward the end of 1945. Lovell's wartime colleague J. S. Hey and his old mentor Patrick Blackett, both also working with ex-radar equipment, disagreed. They both thought that Lovell had witnessed radar echoes from meteors. Lovell's mentor turned out to be right—most of the events were meteor showers—and hope that a new avenue of research on exotic particles quickly evaporated. Lovell countered that meteors were of interest to the astronomical community, noting happily that few astronomical texts paid them much attention and that professional astronomers paid them even less.[12] A novel technique combined with an under-explored area of an established discipline deserved investment. Requiring astronomical knowledge, Lovell began a cooperative arrangement with an amateur meteor astronomer named Manning Prentice, a solicitor by trade. Prentice, presumably enamored by Lovell's novel method of meteor detection, introduced the radio physicist to amateur observational astronomy during the 1946 Perseid meteor showers.

A year after the greatest war ever fought, Lovell and Prentice lay on deck chairs in a field at Jodrell Bank measuring meteor trails and estimating magnitudes of each event.[13] The scene, celebrated by Lovell in his memoirs, strikes us as a couple of retired astronomers engaging in a hobby. The relaxed vista seemingly incongruous against Lovell's self-portrait as being intensely driven toward securing his place in radio astronomy, especially via the Jodrell Bank radio telescope. In fact, the apparent serenity elides the strenuous efforts Lovell was already making to attract astronomers to his work. Lovell and Prentice presented professional astronomers with their observational and radio work later that year and pointedly published in the *Monthly Notices of the Royal Astronomical Society*. Though Lovell's meteor research endeared Jodrell Bank to only a small number of astronomers, he would later claim to have become "part of the astronomical community" after his talk at the Royal Astronomical Society meeting in December of 1946. The Fellows, Lovell later said, "began to grasp that [radar] was a new astronomical technique"; so, evidently, did Lovell.[14]

Lovell's claim to be part of a new community remains among the earliest moments in the creation of the new identity of the radio astronomer. Like Edward Bowen, Lovell avoided any continuing work in air-interception radar or increasing bombing accuracy. We may note how similar their separate stories are when considered side by side. Earlier than Bowen, though, Lovell anxiously tried several early tacks to ingratiate himself with astronomers, including weighing in on a debate over the size of particles that existed permanently in interstellar space. A few professional astronomers, having made crude estimates, hoped that radar could accurately measure the velocities of sporadic meteors, and perhaps even their incident angle, to judge whether meteors possessed enough velocity to orbit the sun. In the end, the question would occupy nearly three years of work at Jodrell Bank, after which it would be concluded that most sporadic meteors do in fact orbit the sun.

Lovell reached outside high-energy physics to the British astronomical community to gain social credit for his work on meteors. His conscious movement between scientific communities was not an isolated occurrence; it was replicated by most future radio astronomers. Making the trip from Australia to Britain in mid 1946, Edward Bowen, for example, made contact with Patrick Blackett to follow up rumors of radar work on cosmic ray showers. Bowen's Australian group had also, like Lovell, tried but failed to detect any exotic particles. Bowen sought priority for the technique of radar detection of cosmic ray showers, offering as proof the results of some 1940 flight tests of early airborne intercept radar equipment. For his part, Lovell emphatically denied that any of the early equipment was capable of detecting cosmic ray ionization. He then announced his plans for a new 218-foot parabolic antenna to be used along with his photographic equipment in "the search for the cosmic ray ionization."[15]

Bowen left England under the distinct impression that the Manchester group held the lead in cosmic ray work involving radio detection. Critically, Lovell's large-instrument response to cosmic rays paralleled Bowen's experience with the American attack on linear acceleration. From that point on, Bowen was convinced that potentially significant fields of postwar research could be secured only by a large investment in instrumentation. As we shall see, Bowen would relentlessly pursue the goal of a giant radio telescope for the southern hemisphere throughout the 1950s, in the wake of Lovell's 218-foot antenna and his 250-foot Jodrell Bank dish. Bowen would thus secure Lovell's parabolic dish antenna as the model for radio telescope design well into the 1960s; the largest incarnation would ultimately be the 1,000-foot Arecibo dish in Costa Rica.

Bowen may have been defeated over cosmic rays, but he had also come armed with news about Pawsey's initial work on identifying sunspots as radio phenomena. Initially, the subject of solar noise interested Lovell only insofar as it was associated with cosmic rays in the upper atmosphere. And so we have a nice illustrative moment when two men's understandings of their scientific community were changing. Bowen didn't think in the same way as Lovell. Though far better connected, especially with the groups stemming from the MIT Radiation Laboratory, Bowen didn't yet conceive of broadening his circle beyond fellow radio physicists. Lovell's and Bowen's changing communities informed both the measure of their radio instruments and the meanings of sunspot noise and meteor echoes. Lovell had approached the astronomical community almost immediately with news about meteors; Bowen traveled by ship for six weeks to speak to fellow radio physicists about sunspots.

Bowen and Lovell exchanged information in 1946 to secure their research areas. Within another year, they both learned that a small band at Cambridge's Cavendish Laboratory was also interested in the phenomenon of solar noise. In fact, from the late 1940s on, one half of radio astronomy emerged from the new social community of radio physics groups at Australia's Radiophysics Laboratory, at Jodrell Bank, and at the Cavendish. As Lovell and Bowen both appreciated, competition justified the elimination areas of research. Instead, the idea slowly dawned that cooperation fostered progress, and indeed a new style of cooperation came into being among three radio physics groups. Lovell "arranged not to overlap" the work of Jodrell Bank and Cambridge.[16] In another instance, Martin Ryle at Cambridge approached Joe Pawsey, Bowen's deputy at Australia's Radiophysics Laboratory, in 1950 with a list of areas that were of interest to his researchers. In particular, he said, "we would be very interested to have your comments, particularly if there seems to be any possibility of overlapping with any projected observations of our own."[17] Pawsey acknowledged some conceivable overlap of research: "What we can do with a very narrow beam aerial in the decimetre range . . . would presumably be in mild competition with your eclipse plan if we proceeded."[18] Pawsey's desire to avoid competition with fellow radio physicists had compelled him to talk through the whole issue with Jack Ratcliffe at the Cavendish back in 1947. Anxious to forge new community ties, the Australians forwarded a list of their own projects as a "means of avoiding possible clashes of interest between our two laboratories."[19]

So we have two themes, ready-made. First, we note that many people used the technology of radar, but three groups took an interest in the detection of ionosphere and celestial noise. Each possessed an avenue of research, which initially shaped how the old radar sets would be used. There was not one path from radar, but many.

Second, and more significant, we note Bowen's desperation as he quickly abandoned linear acceleration and cosmic rays in favor of solar noise but resisted contacting astronomers, while Lovell formulated a plan for a large antenna to detect cosmic rays even while wooing the British astronomical community with talk of meteors. In other words, cooperation and community became part of Bowen's style of science as well as Lovell's, became crucial to the emerging idea of radio astronomy, and became pivotal to the changing notion of what a telescope was. In the years to come, the identity of the radio astronomer would hinge on a community-style social organization in which radio physicists and astronomers would cooperate in crafting new instruments and in training new disciples.

By 1947, Bernard Lovell had secured an observing site at Jodrell Bank, a field outside Manchester, had bought some chairs and a camera, had made an acquaintance with an amateur astronomer, and had attempted to do some science involving radar echoes from meteors. In many ways Lovell had already removed himself physically and intellectually from his institutional base, the physics department at the University of Manchester. No comfortable interdisciplinary culture existed in those old Victorian halls to parallel the modernism of Caltech or MIT.

Still, Lovell gained new colleagues quite quickly. His wartime connections to military sources swelled Jodrell Bank's infrastructure. In and around Manchester, various electrical and electronic business, engineering firms, and suppliers of pipes, metal work, chart recorders, and bakelite insulators fattened Lovell's files with specifications, color brochures, and offers. Some of this advertising paid off—for example, Evershed Chart Recorders received a substantial order for trace paper. Lovell aggressively pursued any contractor that seemed able to build a larger radio antenna, though after more than a year he had not found anyone who could even "guess" at the development costs.[20] The sheer diversity of the industrial effort that the building of the Jodrell Bank facility entailed is worth highlighting, if only because it reminds us of the communal nature of every scientific enterprise, especially astronomy. One of the significant features of big science has been the successful marshalling of industrial and technical resources into a coherent program. Lovell used seven taxi companies to shuttle people to and from the station, four photographic companies, half a dozen construction companies, five valve companies (Priestly and Ford being the most prominent), and about two dozen miscellaneous contractors for everything from ring spanners to linseed oil.[21] Lovell's endeavor at Jodrell Bank didn't yet compare to the giant accelerators already under construction at Berkeley, but Manchester's cosmic ray and radio physics establishment nonetheless required a plethora of wartime contacts and many of Manchester

University's close alliances with local British industries. Those connections would serve Lovell well as he built his giant telescope over the next decade.

Analogous to the diversity of material culture that would help found radio astronomy was the range of human labor and ideas. From very early on, Jodrell Bank had, proportionally, a substantial number of disciples. We know that graduate students do much of the "grunt work" but receive little of the credit. Still, if we can access their work, we can reveal much of the practice of science, especially during those periods of intense ferment of research work before the standard story of discovery takes hold. One issue is the ability of a young and disciplinarily undefined new field of knowledge to produce doctoral and master's students essentially from the outset. Still an understudied aspect of science, students are not the result of academic programs but often the impetus of shifts in academic program. Possessing institutional affiliation and mentorship but not yet invested in professional allegiances, graduate students can more readily move outside disciplines, building the early tenuous bridges between techniques, practices, and practitioners—in short, building community. We can "see" the cultural production of science by looking the lens of graduate student recruitment and training. To that end, I want to draw out the distinct similarities between the new emphasis on graduate instruction at Caltech and Mount Wilson and the experience of Lovell at Manchester. Via a comparison of the experiences of radio astronomy's early graduate students, rather than its research programs, we can begin to see some key elements of the identity and community of the new field of science. Few of the records we have go into Lovell's graduate program in any detail, and the whole effort appears to have been tied to Manchester's own drive for greater prominence through greater numbers of students as well as the larger than usual influx of students into university education in Britain after the war years that saw few students at all. What is clear is that Lovell's graduate students helped create the entire endeavor of radio astronomy. His students performed much of the work in bringing together the wartime radar equipment and exploring early issues of photography, atmospheric phenomenon, and solar radio noise. Institutionally, the Manchester physics department had only two official staff attached to their new research station at Jodrell Bank: Bernard Lovell himself and J. A. Clegg. Yet Lovell successfully funded his core group of graduate students, including four PhD students and two master's students, with a wide variety of grants and scholarships.[22]

Jodrell Bank's graduate students did much of the work of radio astronomy, observing, instrument making, and data collection, but they also formed the key path of the making of the identity of the science itself. In late 1948, Lovell called in his old wartime connection Jack Ratcliffe from Cambridge to help with the oral examinations of his students J. A. Clegg and C. D. Ellyett. As Foucault tells us, graduate students remain illegitimate until they have passed through the rigors of examinations and dissertations. In this instance, Clegg appears to have passed through without incident, but Ellyett seems to have run afoul of Ratcliffe's expectations for a PhD in physics. Though Ratcliffe was nominally a supporter of Jodrell Bank and Lovell, his dissenting opinion of Ellyett's thesis reveals the extent to which discipline is indeed about boundary control and policing. Moreover, it revealed early substantive differences between Ratcliffe's vision of physics and Lovell's of radar astronomy. Ellyett defended his thesis (titled The Photography of Radio Echoes from Meteor Trails) by discussing the irregular time variations his photographs had revealed. In thesis, Lovell and Ratcliffe concurred, Ellyett had "worked out the relevant theory and deduced new and important facts" about the velocities of meteors. But as the oral examination continued, Ratcliffe became increasingly unhappy with Ellyett's knowledge of the "fundamental physics underlying his experiments." Lovell and Ratcliffe asked Ellyett for another essay on his thesis work, and further demanded that he take a written examination on "fundamental physical principles" six months later.[23]

What Ellyett went through over the next six months can only be guessed. A provisional pass on a PhD examination must have been a grave disappointment, and no doubt the coming examination hung like the sword of Damocles. Nonetheless, on June 1 Ellyett sat down to answer his choice of five questions out of eight. The questions reveal much about Ratcliffe's expectations of the fundamental knowledge of a new PhD in physics, and we can see how far Lovell's ideas about radar astronomy had already diverged from Ratcliffe's. We should remember that Ellyett's work concerned photographing meteors and calculating their velocities. We can only speculate as to what sort of exam an astronomy department would have given Ellyett. As it was, Ellyett was required to attempt question two: to explain the power gain of an aerial and to calculate how far from the reflecting surface one should place the detector. The sixth question concerned Ellyett's work directly; it addressed the problem of the velocity of meteorite falling to Earth as a result of gravitational attraction. The remainder of the questions covered various topics in physics. One question was on the kinetic theory of gases, one was on atmospheric pressure, one was on electrons passing through a diatomic gas, and in one the candidate was asked to "deduce an expression for the amplitude of oscillation of a free electron when it is subjected to the field of a radio wave."[24] Though Ellyett had built photographic equipment to photograph meteor event, there wasn't a single question on optics, or one dealing with the astronomical problems such work supposedly entailed. The PhD was duly awarded, though Ellyett's essay impressed Ratcliffe more, he told Lovell, than his performance on the exam.

Four years later Ellyett would have had little trouble. By then, a physicist and an astronomer routinely examined Jodrell Bank's PhD candidates, even though they were still nominally within a physics department. The first astronomer Lovell asked to externally examine one of Jodrell Bank's PhDs was Harry Plaskett of the Oxford Observatory. The thesis topic sounded remarkably like Ellyett's: "the velocities of sporadic meteors." Lovell explained that he was justified in asking Plaskett to help examine the thesis because "although the results have been obtained by radio techniques, the main interest of the thesis [is] astronomical."[25] Plaskett turned Lovell down, citing unfamiliarity with radio astronomy and meteors and a lack of time. "Perhaps on some future occasion, when the examination touches on some branch of ordinary astronomy with which I have at least a nodding acquaintance," Plaskett commented, "you will think of me again."[26] Meteor astronomy was thus neither hill nor mountain, it seemed. It wasn't sufficiently "ordinary" astronomy to interest Plaskett, yet it was sufficiently like astronomy to have Lovell asking astronomers to examine Jodrell Bank PhD candidates. Lovell secured E. Findlay-Freundlich of the St. Andrews Observatory to examine Mary Almond that November.[27]

Becoming a radio astronomer rather than a radio physicist interested in astronomy required renegotiation of the standards of exam questions and of the very conception of each field. The transformation took place quite quickly. By the early 1950s, Manchester's exams contained questions such as the following: "A radio sources [has] an elliptical disc whose apparent angular diameter is 150 seconds of arc along the major axis and 30 seconds of arc along the minor axis. Calculate the equivalent black-body temperature of the disc assuming the intensity to be uniform over the whole disc."[28] That question, unlike any on Ellyett's examination, directly addressed astronomical phenomena, utilizing metrics familiar to astronomers, but asked for results of the kind that a physicist would expect.

By 1957, the year Sputnik was launched, the transformation of an astronomical education had been largely completed. The major changes in curriculum and examinations predated radio astronomy's involvement with the military-industrial complex and didn't follow the traditional stories of new sciences in the Cold War era.[29] Indeed, the process seems the exact opposite, with antenna design, noise reduction, and signal feeds all becoming normalized topics of astronomical training and areas of future research in astronomy in the decade before Sputnik. Nor was this restricted to Manchester University and an indulgent physics department. We find the same phenomenon in a doctoral examination at the Harvard College Observatory, which also supported a doctoral and research program in radio astronomy after 1953, only the other way around. For example, question 4 on the 1954 Harvard astronomy exam went as follows: "At the AAAS Radio Astronomy symposium, Dr. Hagen [a former radar physicist working at the Naval Research Laboratory] suggested that his 50-foot 'dish' might have the greatest available resolution. Discuss this suggestion in view of the large dimensions of some of the interferometric systems."[30]

Thus, less than ten years after the end of World War II, the disciples of astronomy could speak of telescope apertures measured in feet rather than inches. They were expected to understand the principles of interferometers operating at radio wavelengths and to recognize "radio" as one of the tools of their field. At the same time, the radio physicists considered arcs and elliptical discs to be within the purview of their new science. Training regimes changed with the requirements of a changing and broadening community as astronomers rather than physicists were brought in to test new disciples. In short, patronage and instruments were only two facets of the massive changes in science that occurred during the Cold War. In this case, the emerging demands of community reshaped the measure and the meaning of a physics degree in the novel field of radio astronomy.

Perhaps what drew the new community of radio and optical astronomers together was a shared view that students were scarce in the postwar years. The dearth of new blood was, quite simply, "the most important problem facing the United States in the field of astronomy," the new head of Caltech graduate program, Jesse Greenstein, asserted to Peter Van de Kamp, Program Director of Astronomy at the new National Science Foundation. Greenstein insisted that the NSF focus on the "enlargement of the activities of the smaller observatories, and the increase in the number of people involved in astronomical research." In contrast with the "other physical sciences' very great expansions," Greenstein said, he was now witnessing a decrease in the number of departments teaching astronomy at the undergraduate level. He asked whether the NSF couldn't "encourage small universities and colleges to foster teaching, . . . research, [and] fund more modernization of their equipment."[31] In that context, radio astronomy became, in effect, one characteristic of the modernization of astronomy's equipment. American astronomers, Greenstein said, knew full well that "small nations like Holland and Australia" had seen "fantastically rapid growth" in radio astronomy. Yet the United States had contributed only "incidental contributions to electronic circuitry [and] antenna design." Willingness was not lacking; rather, the US simply "lack[ed], nationally, the personnel to carry it through."[32]

Radio astronomy was not a major concern of Caltech's astronomers until 1952. As was noted in chapter 1, Caltech appointed a physicist, Ira Bowen, to direct the new Combined Observatories, a powerful union between Mount Wilson and Palomar that was, moreover, a new alliance between teaching and research. The first postwar annual report of the Carnegie Institute of Washington announced: "The research program of the observatories will be reinforced by studies on the campus of the California Institute and graduate training leading to the doctorate will be given under the auspices of the California Institute by an astrophysics staff consisting of members of both the Institute and the Institution."[33] Rejecting any notion that astronomy was a hand-maiden to physics, Greenstein asserted that Caltech's new foray into pedagogy would "strengthen the relation between astronomy and physics."[34] In the immediate postwar years, this interdisciplinary ethos, itself a feature of Caltech's scientific culture, main-tained an expectation of professionals to train new disciples to combine astronomy and nuclear physics. Rather than distance themselves from what promised to be the most significant scientific topic in history, the optical astronomers at Palomar and Mount Wilson believed that the intersection of astronomy and nuclear physics might reveal new opportunities for astronomy to investigate the properties of stars and the constituency of elemental matter itself. Elements forged in stars through uncertain processes clearly were of as much concern to astronomers as they were to high-temperature, plasma, fluid, and nuclear physicists.

A new ethos of interdisciplinary recruitment and training was forged at Caltech to take advantage of advances in nuclear physics, and would, into the future, aid the radio astronomers at least as much. The larger point here is how the boundaries and identities of scientists crystallize not at the cutting edge of research but in the longue durée of training of new disciples. From the outset, Caltech's president, Lee DuBridge, saw his new graduate program as an "appropriate accompaniment to the great program of observational astronomy."[35] Caltech expected a combination of transdis-ciplinary research, superior instruments, and new opportunities for institutional coop-eration and exchange to yield substantial dividends in students. "Most of the very best students will be strongly attracted by the opportunity of contact with the Mount Wilson group, and with the prospect of use of the telescopes," Greenstein anticipated.[36] As it turned out, Greenstein fondly remembered in later years that "students [had] pretty much unrestricted access to the 100-inch," which he characterized as "a good telescope for a student."[37]

Caltech's new graduate program in astronomy had profound implications for other American observatories, especially those in decline. The Harvard College Observa-tory's director, Harlow Shapley, believed—or perhaps desperately hoped—that the

Caltech program would "coordinate nicely with the programs at the Harvard, Princeton, and Yerkes Observatories."[38] DuBridge assured Shapley that although Caltech would conduct graduate education as well as "post-doctoral research training," the program would emphasize only "certain fields, for example, astrophysics, which we are uniquely equipped to undertake and which may not conflict with programs in other centers."[39] Disciplinary competition wouldn't pit one institution against another; rather, cooperation, exchange, and flow would characterize the new community of knowledge. Clearly not convinced, Bart Bok, another Harvard astronomer, wanted DuBridge to "drop around" when he was next on the East Coast and talk about the new Caltech program, no doubt seeking reassurance that cooperation really was Caltech's intention.[40]

Concerned that any advertising of the new astronomy program would create a flood of transfers from graduate students at other institutions, one Caltech dean warned Greenstein that he "should be very careful not to give the impression to other astronomy groups that we are attempting to proselyte students who have already begun work elsewhere."[41] Greenstein worried that Caltech's new program would attract "too large a concentration of good students of physics and mathematics, with little observational interest or experience on 'deep' theory."[42] Evidently, superior skills in physics or mathematics didn't immediately equate in Greenstein's mind with better graduate students of astronomy. Caltech and the Combined Observatories sought students with a distinct bent toward instrumentation and, preferably, some observing experience.

Parallel to the experience of Manchester, Caltech's development in the years after World War II exposes how tightly sciences are tied to their cachet among potential new practitioners. The expansion of the astronomical community began with high expectations for the merger between Caltech and Mount Wilson. Integration of the staffs of the two institutions took place rapidly, evidently (aside from Edwin Hubble) without any ill will. Jesse Greenstein felt particular relief that the two staffs cooperated on undergraduate and graduate teaching. Indeed, the most immediate and significant benefit of the merger was the influx of multi-disciplinary teachers into astronomy. "Robert King has moved down to Caltech from Mt. Wilson, into the physics department, to teach optics and build up spectroscopy," Greenstein told Bart Bok at Harvard. "Oliver Wulf is here [and] some of the chemists also have astronomical leaning[.] Tolman's death was a loss, but H. P. Robertson is starting a special relativity course, and will be a great help."[43] In later years, Greenstein would also remark on the informal side of a graduate education. Several people, he remembered, though not officially lecturing, would turn up at Caltech and just "talk to people" in the astronomy department. Walter Baade and Rudolph Minkowski, Greenstein reminisced, were interested in everything, especially "radio astronomy and supernovae problems."[44] Teething problems abounded, of course. Few of the staff members of the observatories had ever taught, and the instructors supplied by the Carnegie Institution were very young and inexperienced. Caltech's early curriculum in astronomy was rather haphazard. In his first year, 1948, Greenstein was required only to give a single whole-year course on "'astrophysics' (whatever that means)," he commented to Bok. Greenstein's uncertainty did not affect enrollment though, he boasted to his old advisor at Harvard, as "8 registered for the course, [and] 17 attended the first lecture" yet the department had only five graduate students. It seemed a promising start to the entire endeavor. Still, he was in no doubt, he joked, that "the bleak days when two appear, one of them a loud sleeper, are undoubtedly ahead of me."[45]

Greenstein's humorous optimism about his future undergraduates soon dwindled. Both Greenstein and Ira Bowen anticipated relying on their institution's superior observing facilities to successfully attract students but, in fact, in the years following the end of the war Caltech astronomy barely grew while other sciences exploded. For instance, in the mid 1950s Caltech's astronomy department had only two NSF-funded postdocs, while its mathematics department had 16 and its physics department 19. And expansion of the staff outpaced the increase in the number of graduate students. By the mid 1950s, physics accounted for 62 of the 89 staff members, while mathematics had only 18 and astronomy only 9. But why? Graduate education at Caltech and Mount Wilson could hardly have been more idyllic. Walter Baade personally instructed Allan Sandage, Caltech's first graduate student in astronomy, how to use the big telescope on Mount Wilson.[46] Sandage sat "night after night on Mount Wilson when Baade was observing, listening to Baade talk." Sandage's education, Greenstein noted years later, was "a little bit better than a one year course on math physics" that Caltech students were receiving by the 1970s.[47] And clearly Bowen's and Greenstein's downcast evaluation was not entirely borne out. Sandage, who showed an extraordinary desire to master physics and became an adept user of the 200-inch telescope, was exactly the sort of astronomer that Caltech had hoped to train when the graduate program was established. By 1953, when Edwin Hubble became ill and died, Sandage had already outlined plans to pursue Hubble's elusive constant. That work would keep him busy for about 30 years and would earn him recognition as one of the first cosmologists.[48]

Student recruitment and training, then, helped shape the interdisciplinary radio astronomy community in both the United States and Britain. While Manchester and Caltech emphasized their efforts in crossing the boundaries between disciplines, transforming those national efforts into a broader international community would take the Australians, who played the role of matchmaker. As Edward Bowen sailed home from Britain to Australia in early 1947, in Sydney his staff continued to investigate the patterns and occurrence of solar radio noise and electron acceleration and to work on civilian air traffic control. Just as Bowen reached Sydney, Australia's linear accelerator program folded, but the group working on civilian air traffic control had begun promising field tests of their navigation and guidance system, known as the Multiple Track Range. Meanwhile, Joe Pawsey's group continued to make detailed recordings of solar noise levels at a research station at Dover Heights on the picturesque South Head of Sydney Harbor.

Once more, to appreciate why radio astronomy emerged as a community-style science during the Cold War we need to consider the alternatives, in this case the seemingly logical step for a radar research laboratory to work on air traffic control systems. As war work wound down in early 1945, applying wartime radar technology to civil aviation operations and safety seemed an obvious and worthwhile avenue for radio physicists. Bowen personally witnessed the effects of primitive radar navigation and the resulting numerous casualties (the subject of one of Arthur C. Clarke's earliest postwar novels, *Glide Path*). Within the immediate postwar program of Australia's Radiophysics Laboratory, "radar aids to civil aviation" comprised two projects already in development by war's end: the "Distance Measuring Equipment," which gave the distance between the ground and the aircraft, and "Multiple Track Range," which generated multiple radio paths defining air routes around an airport. Results seemed easy, and the road ahead clear. The projects coalesced into workable systems very quickly. With obvious relevance to the idea of turning swords into plowshares, Bowen described radar for civil aviation in the *Australian Aviation Annual* as "perhaps the most important applied work" of the Radiophysics Laboratory.[49] Late in 1946, the Australian Department of Civil Aviation agreed to full-scale trials of the Multiple Track Range on the Sydney-to-Melbourne air route.[50] Australia's CSIR was not alone in believing that developing international radar systems for civil aviation was precisely the sort of thing that civilian postwar science should investigate.[51] In the years 1944–1947, standardization, international agreements, and a system of global technical networks were the subjects of a series of Commonwealth and Empire Conferences on Radio for Civil Aviation. This international cooperative allocated specific research programs to the various members of the British Commonwealth on the basis of their experience in radar and their research resources.

Australia's assignment to work on Distance Measuring Equipment presumably stemmed from Edward Bowen's wartime work on radar systems for air interception. DME straightforwardly told a pilot his aircraft's distance from a fixed beacon. The Australian system had undergone flight tests in July of 1945. Likewise, the August 1945 field trials of Australia's Multiple Track Range were definitively successful, and conversations during Bowen's trip to Britain in late 1946 apparently encouraged him to continue MTR's development. Bowen arrived back in Sydney under the impression that British "experimental establishments are in very bad shape" and considered the Radiophysics Laboratory's work "surprisingly far ahead of England [both] in appreciation of the problem and in arriving at a solution" for civil aviation.[52]

A year later, though successful tests had been performed, the Australian MTR system was not recognized for international use, and Bowen rapidly wound down the group working on the prototype. The honor went instead to Britain's Telecommunications Research Establishment, while the Canadian DME gained international favor. The Australians left the field bitterly disappointed. Their former Scientific Liaison Officer attached to the Radiation Laboratory at MIT, V. D. Burgmann, thought the Radiophysics Laboratory's system superior, since the others "do not possess the extreme simplicity which characterises the Australian MTR." In a seeming repetition of the linear accelerator program, Bowen's superior at the CSIRO, Frederick White, once again cautioned Bowen about the "enormous effort which the Americans have now transferred to this field of endeavour. [They] are trying everything and while 80 per cent of it can be discounted out of hand, they are bound to arrive at some excellent solutions if only by process of elimination."[53] Still, in a final twist, Australia's civil aviation authorities independently continued development of the Radiophysics Laboratory's Distance Measuring Equipment. Indeed, the Radiophysics Laboratory's DME was first introduced on a commercial service in 1950, and by 1954 it had become mandatory on all Australian civil aircraft.[54]

The aforementioned incident may have been a minor one in the history of Cold War–era science, but it is illuminating nonetheless. Postwar Australian work in radio physics remained situated within a larger set of British Commonwealth institutions competing for honors. In turn, those Commonwealth organizations neighbored the growing American research endeavors engaged in exactly the same developmental work. Still, the larger point here is not who got laurels but the immediate postwar character of the radio physics community. In the series of international competitions and conferences, the global network of radio physics expertise remained at work, directed toward specific problems of direct relevance to the nation-state. Though national efforts competed, open presentations and demonstrations decided the result; the networks of practitioners left over from the war remained intact and operational. Still, for those at Edward Bowen's Radiophysics Laboratory the defeat of both their DME and MTR in the international arena was a harsh blow. It also left essentially only one major subject of research: solar noise.

Like the ignominious defeat of the Australian linear accelerator before the vastness of the American research-and-development resources, the international rejection of Australian radar aids to civil aviation made the creation of radio astronomy possible. Without situating Joe Pawsey's solar noise group in a context of promising, alternative, but ultimately frustrated trajectories for wartime radar technology and expertise, we cannot appreciate the measure and meaning of the new radio astronomy in the late 1940s and into the 1950s—in short, radio physicists' embrace of a lack of application to national problems as well as their commitment to an international and interdisciplinary community.

Pawsey's solar noise team expanded as the linear accelerator and radar aids for civil aviation projects faltered. Solar noise's promise rested on its unexpected ability to cross both national and disciplinary borders. Edward Bowen regarded radio noise as "clearly of fundamental importance to astrophysics since it provides a method of investigating extra terrestrial phenomena by means other than light." Radio techniques became understood as existing outside national and disciplinary boundaries, as well as holding appeal precisely because of their non-applicability. For Bowen and Pawsey, "success already achieved gives strong grounds for hoping that important astrophysical discoveries relating to the structure of the outer atmosphere of the sun and stars will derive from further study."[55] Solar noise would leave the mundane world of particle acceleration and civilian air traffic control behind and instead would gaze at, and understand, the celestial heavens.

To pursue this now-valued research field, Australia's Radiophysics Laboratory required people. Recruitment and training were central to the transformation of the Australian radio physicists into radio astronomers, as elsewhere, but the process differed from that at Jodrell Bank. Until the later 1940s, Bowen and Pawsey could internally transfer interested or usefully qualified people to other parallel projects with little trouble. Thus, work on solar noise and (later) work on cosmic noise benefited from the death of the Australian linear accelerator and the defeat of Australian ambitions in radar aids for civilian aviation. A distinct feature of the Radiophysics Laboratory's organization remained that the Australians possessed little opportunity to train students via an associated graduate program. Fully qualified people would have to be lured to Sydney; but who, in 1946 or 1947, could possibly have been qualified in a field that was still known as "noise"? While Bernard Lovell's efforts near Manchester existed within an institutional context of a university, the Radiophysics Laboratory existed apart from the Australian university system. Consequently, recruitment operated very differently in Australia. While university-based programs could recruit and train new disciples in their own image, Australia's inability to internally generate disciples forced the Bowen and Pawsey to establish community wherever they could.

The best example of the indeterminate nature of Australia recruitment and training was John Gatenby Bolton. Having received a basic science degree from Cambridge in the late 1930s before enlisting in the Royal Navy as a radar operator, Bolton responded to Bowen's advertisement for researchers in mid 1946 and began working within Pawsey's solar noise group later that same year. Pawsey and Bolton were not altogether amiable colleagues, as it turned out. Bolton's impetuous rush toward new discoveries and his self-conscious promotion placed him at odds with the shy, patient, methodical Pawsey. On the other hand, Bolton and Bowen were made for each other, both being committed to a vision of radio astronomy that was big, bold, and brash. An ambitious, driven young man, John Bolton argued that galactic noise could rise above the cacophony produced by the sun. He attacked Pawsey's research program; he also challenged the theoretical basis of "galactic noise" that Karl Jansky had proposed in the 1930s.[56] Bolton suggested that noise could be pinpointed to individual sources, and that the evidently homogeneous distribution of noise across the galaxy could be merely the summation of the individual sources. After turning the solar group's antennas to the constellation Cygnus against Pawsey's instructions, Bolton, with his colleagues Bruce Slee and Gordon Stanley, detected the first extragalactic radio source, Cygnus A. The discovery of a number of radio sources enhanced Bolton's scientific reputation, as did the first correlation between a radio source and an optical object (the Crab Nebula), which he achieved in 1948.[57]

The discovery of the Crab Nebula as a radio source was a transformative moment for radio astronomy. That nebula had been a focus of astronomical investigation since the tenth century, when it went nova. The discovery that it also emitted radio noise astonished the optical astronomical community. The foremost Dutch astronomer, Jan Oort, already extremely interested in the possibilities of radio techniques, wrote Bolton a long letter describing current astronomical work on the Crab.[58] Oort's support generated significant interest among the wider optical astronomical community for the first time, particularly in the United States. Likewise, during a trip to the United States and Canada in 1949, Pawsey noted that astronomers were now "excited" by Bolton's earlier identification of the Cygnus point source and agreed that it was "a new astronomical entity."[59] Bolton later expressed a feeling that the identification of a radio source in the Crab Nebula had gained him "respectability as far as the 'conventional' astronomer was concerned."[60] In the late 1940s, no astronomical institution had begun any developmental work in radio astronomy, although astronomers' interest had been certainly been piqued by Australian and British revelations. "As a result the Australian work," Pawsey reported home, the "position now is that the astronomers of the US [have] become thoroughly interested in the implications but have not yet taken the plunge of tackling a totally new technique. Meanwhile, the physicists [now] have other interests. The result is that we have a first class opportunity to establish the lead which we at present hold."[61]

Joe Pawsey's trip around the United States and Canada in 1949 made even deeper inroads for radio physicists into the existing astronomical community than Lovell's amateurish attempts. Through Bowen's wartime friendship with the new president of Caltech, Lee DuBridge, Pawsey readily got audiences at Mount Wilson. While the "astronomical center of the world" excitedly quizzed Pawsey about the details of new radio sources, Lovell's meteor work didn't resonate nearly as strongly with the astronomical community. Throughout 1946, 1947, and 1948, English astronomers had maintained polite interest in Lovell's radar echoes from meteors studies but had rarely gone beyond formality. Still, the whole exercise ultimately concerned community building. Pawsey's reports back to Sydney indicate the type of observations he was making when visiting American laboratories. His tables of institutions, their people, and their projects all detailed the directions they were taking. Lectures, informal seminars, and conversations with the head of the new Caltech graduate program, Jesse Greenstein, or with that doyen of optical astronomy Rudolph Minkowski atop Mount Wilson, enabled information to flow between the dynamic new group of would-be astronomers and their better-established future colleagues.

The friendly and easy rapport so readily established between Caltech and Mount Wilson and Australia's Radiophysics Laboratory would be one of the foundations of the radio astronomy community. When, in 1950, Bowen applied to the CSIRO for another extended trip for one of his researchers (this time for John Bolton), he anticipated another act of community building. Bolton furthered the Radiophysics Lab's contacts in the American astronomical community, and continually emphasized the relevance and significance of radio work to his primarily optical audiences. Such efforts at community building were not based solely on idealism or utopian visions. For the Australians, at least, a substantial pragmatic and self-interested motive lay behind their charitable worldwide speaking engagements. "There is," Bowen wrote, "no doubt that CSIRO work is held in high regard in the USA and a not undesirable by-product of Bolton's visit would certainly be further 'publicity' of the right kind" for Australian radio astronomy.[62] Still, the process of community building went in both directions with welcoming, indeed almost overwhelming, receptions to a representative of a small Australian radio physics laboratory trying to do astronomy. Bowen, Pawsey, and Bolton would work tirelessly and travel endlessly for the next decade to bind radio and optical expertise into a single community. Likewise, Bowen and Bolton would eventually unite two hemispheres together under a single radio sky when Bowen opened his Parkes radio telescope and Bolton his Owens Valley Radio Observatory.

Cooperation rapidly yielded significant insights into such foundational questions as the shape of the galaxy. John Bolton and Ken Westfold charted the direction of the winding of the spiral arms of the galaxy via radio, while Greenstein pursued the same question optically from Palomar Mountain. The research crossed disciplinary lines easily, the Australian radio astronomers' letter appearing *Nature* in 1950 and Greenstein's abstract on the topic in the *Astronomical Journal* in 1949. Not only did all the participants correspond openly; they exchanged reprints and data rapidly and hospitably. Likewise, astronomers shared theoretical speculations about radio data that the radiophysicists had been amassing since the 1940s. One theory, popular in the early 1950s, held that the power of the radio emissions from celestial objects came from "free-free" thermal transitions in electrons. In the case of the Crab Nebula, however, Greenstein's own work dismissed that idea: "It appears very probable that the Crab Nebula actually has a very high electron temperature; even at 200,000° we find that the nebula becomes optically thick at about 100 megacycles. In that case the predicted flux varies as the frequency squared at low frequencies and becomes constant at high frequencies. Thus, both in absolute magnitude and intensity distribution, the free-free emission hypothesis must fail."[63] What better sign of the new nature of astronomy could there be than Greenstein's citing radio astronomy data on the temperature at radio wavelengths as easily as he cited optical absolute-magnitude work? Over the next decade, both Rudolph Minkowski and Walter Baade would devote substantial use of the 200-inch telescope to the identification of radio sources and many long nights to the task of correlating radio and optical objects.

In 1949 the Soviet Union exploded an atomic bomb and reasserted its weight in the emerging struggle between the two superpowers. North Korea, supported by the Soviet Union and now-communist China, attempted to reunite the peninsula forcibly and was held off only by a series of audacious amphibious landings by United Nations forces under General Douglas Macarthur. The Cold War had begun in earnest.

In comparison with such an epic tale the stagnation in the numbers of students taking up radio astronomy may seem a small story; however, it humanizes the vast social changes the Cold War had already unleashed. In short, the Cold War had already channeled students of science and engineering into the concerns of the new permanently mobilized military state. Graduate recruits of poorer quality seemed all that was left for fields of pure science, astronomers bemoaned, at least in the United States. In contrast with the experience of his friend at the Leiden Observatory, Henk van de Hulst, Jesse Greenstein had found that Caltech students were "not 'successful, rich, [or] hard-working.'" Greenstein, having been brought to Caltech to invigorate the institute's new graduate program, lamented how such "boys do not anywhere in the world try and become astronomers." Now, owing to the glamour of nuclear physics and electronics, the applicants that astronomy received were even less promising. "We have only a few students that are good, we hope to have more," Greenstein told van de Hulst. But the creation of the new graduate school and the new emphasis on instruction on the part of the staff at Caltech and the research staff atop the mountains signaled an increase in the moral capital attached to astronomy's ability to recruit and train new disciples. Greenstein's early impressions didn't inspire confidence, however. "They are the same boys you know," he said to van de Hulst, "lazy or hard-working, unhappy and uncertain of themselves. They follow a distant unattainable dream."[64] Might the new radio astronomy have been understood as a salve to the collective ennui of men of science in the early days of the Cold War in the face of the military-industrial complex?

3 DISCIPLES

The art of the proper training of graduate students is not an easy one, but I think honestly that I cannot think of anything more rewarding to do in this world.[1]

—Bart Bok, Harvard University, 1954

As a young radio physicist, Robert Hanbury Brown spent World War II working in drafty airplane hangars and going up and down in planes testing various aspects of airborne radar. After the war, he returned to Britain's Telecommunications Research Establishment before following the father of British radar, Robert Watson-Watt, to Canada to work for Watson-Watt's private scientific consulting company. He promptly became an early casualty of the failure of that entrepreneurial endeavor. Unemployed, he returned to Britain and was recommended to Bernard Lovell's growing group at Jodrell Bank by Patrick Blackett, a professor of physics at Manchester. Lovell remembered Hanbury Brown as a fellow radar "boffin" from the war days. As near contemporaries, Hanbury Brown and Lovell shared parallel careers within the British wartime scientific establishment as "boffins," a charming English term denoting inventive skill drawn from tacit knowledge and some pluck.[2] Lovell immediately offered Hanbury Brown a position in Manchester's cosmic noise program. Though now a radio astronomer, Hanbury Brown readily admitted (indeed fondly recalled) that he "didn't know any astronomy" but "reckoned that it would be easy to learn enough as I went along."[3]

Hanbury Brown entered the University of Manchester as a PhD student in 1949. His working knowledge of radar technology helped legitimize the radio astronomer as a new type of astronomer whose expertise lay in circuits and antennas, not telescopes and spectroscopes. The figure of Hanbury Brown also focuses our attention on the extent to which Lovell's endeavor at Jodrell Bank, among others, grew on the backs of graduate students. Recovering the role of the student in the formulation of new knowledge owes much to the pathbreaking work of Andrew Warwick. In *Masters of Theory*, Warwick explains how the spread of General Relativity in the 1920s, like that of Maxwell's electromagnetic theory 50 years earlier, took place largely in the classrooms of Cambridge. Both in content and in style, pedagogy supplies much of the formulation, practice, and dissemination of science. As Warwick revealed, both James Clerk Maxwell and later Arthur Eddington relied on the intense undergraduate mathematical training regime at Cambridge to fully explore, explicate, and defend their novel conceptions of nature. The performance of the classroom acted out the defense of the theory. In their Senate House exams, however, the brightest students were expected to challenge their master's proofs and taught techniques, and even to generate novel results.[4] For the radio astronomers, pedagogy taught the new disciples how to use radio telescopes and how to become astronomers. Hanbury Brown's arrival at Jodrell Bank changed the very object of study from meteors to galactic and extra-galactic radio sources, and the techniques of studying those objects from intermediary photographic and radio techniques to using large parabolic dishes.

The previous two chapters established that Bernard Lovell, Edward Bowen, Martin Ryle, and several American astronomers really had no coherent idea of what radio astronomy might be in the late 1940s. In that tumultuous period, the leader of Manchester's radio astronomy program encouraged a wide variety of projects in order to build up equipment and numbers, betting that a program's size ultimately mattered in the institutional context of postwar Britain. It was a pattern of institutional and instrument expansion replicated in Cambridge and in Australia in the early 1950s. This chapter examines not only the expansion of the radio physicists into the world of the astronomers, but also the simultaneous and equally significant transformation of the optical astronomers to see radio telescopes and radio physicists included in their science. The first half of the chapter deals with the former change at Manchester and then in Australia. The second half traces the latter changes at the Harvard College Observatory and at Caltech. In all these cases, the character of radio astronomy came from the simultaneous construction of the radio telescope via radio astronomy's disciples and also from the construction of radio astronomy's new disciples via radio telescopes. In other words, radio astronomy's disciples and its telescopes were dependent on each other for their measure and their meaning.

In 1949, when Lovell was studying meteors with nothing more than some old radar sets, a camera, and a few deck chairs, he had a graduate student begin a project to build a large parabolic dish. The dish, built entirely of wire strung from scaffolding tubes, eventually reached a diameter of 218 feet and a depth of 24 feet. It was termed the "transit telescope" because the Earth's motion effectively caused the heavens to transit across the beam. At that size, the dish was an order of magnitude bigger than any other antenna then in existence. Lovell's planned cosmic ray experiments necessitated a substantial increase in receiver sensitivity, which dictated the instrument's size. The dish offered the added simplicity of having only one dipole antenna system at the focus as opposed to a broadside array in which a large number of dipoles and connections would have to be altered each time researchers wanted to change wavelengths. Lovell later reasoned that this simplicity outweighed the difficulty of constructing a wire parabola.[5] Once again, Lovell utilized his wartime connections, "borrowing" some of the newest radar receivers from J. S. Hey at the Army Research Group on the eve of Robert Hanbury Brown's arrival at Manchester. Armed with more sensitive receivers, Lovell disconcertedly learned (from yet another graduate student, Victor Hughes) that the limit of noise for the telescope pointing at the sky was now set by the unexplained phenomenon of "cosmic noise" that Karl Jansky had happened upon in the 1930s and that John Bolton in Australia was attempting to isolate to a particular astronomical object. Soon after his arrival, Hanbury Brown turned the transit telescope toward the problem of cosmic noise and abandoned Lovell's plan for research into cosmic rays.

The transit telescope took Hanbury Brown nearly a year to bring into full operation. Given the directional confines of the instrument (initially limited to a single declination, 53°N), he concentrated first on the Andromeda nebula, taking 90 days in mid 1950 to produce a map of the nebula measuring 1 hour by 6 degrees. It took Hanbury Brown's considerable talents to learn how to adjust the central 126-foot mast of the telescope so it could scan other parts of the sky. "To avoid kinking the mast," he later recalled, "we had to tilt it in almost imperceptible stages[,] and on a good day, when it wasn't raining or snowing, it took us about two hours running from guy to guy shouting at each other and peering though theodolites, to move the beam through one-beam-width (2 degrees)." Here is a classic "boffin" image of a young researcher struggling to make equipment work in the face of that very English adversary, the weather. Yet he was also struggling to utilize and define a new type of instrument, one that had familiar components but distinctly unfamiliar purposes. He wanted to scan more of the sky, but why he might want to do so, or what he might find there, remained a mystery. Of course, the image of running around a muddy field became romanticized in his memory. After leading a large astronomical facility in Australia in later life, he commented: "When nowadays I see people in nice warm control rooms drinking coffee and swinging the beams of their telescopes about the sky by simply pushing buttons, I think of the hours and hours we spent steering the beam of that telescope."[6]

All Hanbury Brown knew in 1950 was the sheer effort it took to point his instrument at other parts of the sky. (By 1980, however, the instrument could be commanded to go to specific points of interest at will.) Radio astronomers—all of them graduate students—struggled with their instruments in the same way they struggled to become understood as part of the broader astronomical community. In the two decades after 1945 they ran from wire to wire, fiddling, tinkering, fixing, and hoping; a generation later, these students controlled a radio astronomy community spanning half a dozen countries, commanding a dozen major instruments, and uniting at least three disciplines: astronomy, radio physics, and electrical engineering. In short, it took a generation for the disciples to become masters.

Hanbury Brown's map of the Andromeda galaxy began a broad survey project, which eventually charted 28 galaxies that looked normal when observed with optical telescopes. His radio map of Andromeda attracted the interest of local Manchester astronomer Zdeněk Kopal, who suggested that the supernova remnant of Tycho Brahe's new star of 1572 might prove to be a radio source.[7] Hanbury Brown shared his radio surveys with Kopal; Kopal offered encouragement and potential direction based on his astronomical expertise. Thus, through cooperation and interdisciplinary exchange, Hanbury Brown and Kopal attacked one half of radio astronomy's main question of the 1950s: Were radio sources "normal" astronomical objects, or were they exotic and perhaps extra-astronomical objects? After trying to ingratiate himself with astronomers through his work on meteors (objects perhaps so normal as to be boring), Lovell eagerly embraced the attention his motley band of graduate students and instruments now attracted. Kopal, who in the 1950s visited Jodrell Bank every week to deliver lectures, "tried to teach [Lovell, Hanbury Brown, and their graduate students] astronomy." Meanwhile, Patrick Blackett charged Kopal with making the Jodrell Bank group "astronomy-minded."[8]

The transformation of the radio physicist into the radio astronomer, and of the radio antenna into the radio telescope, took place rapidly. In fact, although it took two weeks of preparation, it actually occurred in the course of a weekend. If we consider the action slowly, we can see the monumental changes Lovell and Hanbury Brown made in the idea of astronomy. In July of 1949, some sixty Fellows of the Royal Astronomical Society traveled to Manchester for a two-day meeting and an informational tour of Jodrell Bank. It was only the second time the society had met outside London. Lovell proudly informed his Jodrell Bank colleagues that the honor couldn't be of "greater significance to Jodrell Bank and to all of us individually." Thus, they all "must do their utmost" to ensure the success of the Royal Astronomical Society's visit. Radio work on astronomical topics had secured Jodrell Bank's practitioners their "magnificent accommodation." A response not "worthy" of it could see it stripped ingloriously from them all, and all their efforts "shameful[ly]" abandoned. "Search out" your "best photographs or results," Lovell instructed, "and get them reproduced or platted on a grand scale."

Impression and image were most important. Lovell ordered charts and data turned into pictures. "Remember our visitors will be experts in astronomy, but will probably not know much about radio. . . . Prepare what you are going to say accordingly—simple explanation of technique, but not of meteors or sunspots!"[9] This was the moment radio physics became radio astronomy. The Royal Astronomical Society came to Lovell, but Lovell made sure that his visitors recognized that photographs took precedence over mere data. Lovell and the society merged the social and intellectual strands of a scientific community. Jodrell Bank combined radio techniques with familiar astronomical presentations and gave priority to astronomical relevance.

Retracing Lovell's careful preparations, we can witness the process by which the idea of a science changes through the building of new communities. Lovell insisted that the entire weekend be a spectacular performance. "Many visitors complain that our showmanship is abysmal. Few of you can tell a good straightforward story of what you are doing, or readily show convincing results without searching through acres of paper or miles of film."[10] The sociologist of science Bruno Latour once argued that the popular understanding of science was as "a body of practices widely regarded by outsiders as well organized, logical, and coherent." "In fact," Latour observed, science "consists of a disordered array of observations with which scientists struggle to produce order."[11] Lovell and his graduate students imposed a Latourian order on the disordered mess of radio charts and data, building a picture of the radio astronomy instrument as well as the radio astronomer for their astronomical audience. The Jodrell Bank group's performance ordered radio astronomy within the body of astronomical knowledge. It also demonstrated how the radio astronomy community would work to bring radio and optical people into a coherent social community.

That work fell, once more, largely to the graduate students. Ellyett and Greenhow collaged the meteor velocity photographs. Moran and Gatenby assembled diagrams of moon echoes including an "apparatus functioning." Hughes and Little plotted Jodrell's Cygnus runs and mentioned Cambridge's results, while Clegg (as the only other staff member) and Closs handled the theoretical description of antennas. Lovell himself took on the formidable task of selling the "Large Jodrell Plan."[12] All papers had been read aloud two weeks before, and all photographs mounted and put on display a week before, the official "target date." Display, authority, and organization governed Lovell's quasi-militaristic planning for the Royal Astronomical Society's

Fellows. But in that organizational mode, the operation of radio astronomy community at Jodrell Bank became clearly visible to outsiders, as clearly visible as radio astronomy knowledge became to the visiting astronomers by way of photographs.[13] The graduate students generally handled instruments and data, while Lovell handled the synthetic big picture and the bigger plans.

After two days of wining and dining, lectures, discussions, tours, and big plans, Lovell unveiled before the Royal Astronomical Society his ambitious and wildly expensive plan for a giant radio telescope. Still, the astronomers didn't fully know what to make of Lovell's performance. At the last dinner, the Society's president "congratulated the Jodrell Bank team on this new development in observational astronomy (or physics)."[14] The literal text of the speech betrayed optical astronomers' continued uncertainty about the exact nature of Lovell's program. It may be the best description of early radio astronomy we have: "observational astronomy (or physics)." Radio astronomy sat on the knife-edge, poised between the world of physics and the world of astronomy. On one point, however, the Royal Astronomical Society was clear: "The high-spot of the meeting was undoubtedly the show you put on at Jodrell Bank."[15] The show was the first step in the long process whereby radio physics became radio astronomy. Lovell, Clegg, Ellyett, and Moran crafted their messages to be read in astronomical terms. Like astronomers the world over, the Fellows accepted the importance of instrumental prowess to scientific status and so enthusiastically listened to Lovell's long speeches and grandiose plans for a giant radio telescope in Britain. As Englishmen, they understood Britain's declining state; as members of a technocratic elite, they were impressed by technical innovations as an approach to solving scientific problems. To top it all, they witnessed a new team of disciples hard at work effectively transforming radio charts and data into comprehensible all-wave astronomy.

In January of 1950, as a direct result of the success of the Manchester meeting, the Royal Astronomical Society formed a Committee on Radio Astronomy. This group of astronomers and radio astronomers aimed to "keep radio workers who are exploring new methods in touch with astronomers of the classical type who might be able to help by suggesting problems for investigation."[16] Patrick Blackett, head of the Department of Physics at Manchester and Lovell's boss, initiated the Committee as one "link" between radio astronomy and "classical" astronomy. Blackett believed that the committee could assist in supporting and promoting whichever "research projects would best serve to maintain the lead which this country had gained in Radio Astronomy."[17] Moreover, the Society threw its weight behind Manchester's big dish a year after Lovell's star performance, lauding a new astronomy that was "independent of climatic conditions."[18] The steel-gray skies of Manchester proved no obstacle to radio's vision—a lesson learned well during the Battle of Britain when radar saw through the clouds to detect German bombers.

We should also note that radio physics changed into radio astronomy between 1948 and mid 1950 because Lovell and his graduate students became more concerned with the "astronomical significance" of their work than with "fundamental physical principles." In part, one could cynically argue that Lovell simply responded to the market: that with nuclear physics already sapping most of the funding opportunities from physics, he merely crept ever closer to astronomy. Indeed, Lovell's first big grant was from the Department of Scientific and Industrial Research, which had also been impressed by the "timeliness and promise of the investigations of meteors, the upper atmosphere, and other extra-terrestrial phenomena by radio technique."[19] With one grant in hand, Lovell rationally pursued likely patrons. Yet we must appreciate that funding is not the sole concern of a scientist, nor was potentially fruitful scientific direction obvious in the confused world of the early postwar world. Radio astronomy's young boffins, its new pictures, and its instruments didn't yet form a coherent identity, as others recognized. Jack Ratcliffe at Cambridge (an older man) saw Lovell's work in a tradition of ionospheric physics; Lovell himself came to view radio physics as a technique of astronomy; the president of the Royal Astronomical Society genuinely didn't know what it was.

Though the trajectory of radio physicists wandering in search of another community looked similar in Australia and in Britain, the initial reception of the two groups was remarkably different. As we have seen, after a period of bemused indifference, Lovell eventually made the Royal Astronomical Society sit up and take notice of his Jodrell Bank work by literally inscribing radio charts into visible pictures and by ruthlessly controlling the content of his staff's presentations to ensure that they sounded astronomical and not radio-physical. In Australia, by contrast, Edward Bowen and Joseph Lade Pawsey of the Commonwealth Scientific and Industrial Research Organization's Division of Radiophysics initially got a quite hostile and antagonistic reaction to their entire endeavor from Commonwealth Astronomer Richard Woolley.

During the war both the CSIRO's Division of Radiophysics (headquartered at the University of Sydney) and the Commonwealth Observatory (on Mount Stromlo, near Canberra) pursued investigations of radio disturbances in the upper atmosphere. Weak linkages between the two organizations continued after the war, largely driven by Joseph Pawsey's need for optical solar observations to correlate with his radio graphs of solar noise. But in 1946, when Bowen and Pawsey asked Woolley for a far more systematic and detailed optical recordings of the visible sun, Woolley rejected out of hand any suggestion that his observatory should devote resources to charting radio noise.

The stormy relationship between Woolley and Bowen (who seemed to have utterly disliked one another) probably provided an impetus for Bowen to re-connect with his friends from the MIT Radiation Laboratory—particularly Lee DuBridge, now the president of Caltech, which had just completed merger negotiations with the Carnegie Institution's Mount Wilson Observatory. At the same moment when Bowen's news of Australian solar radio noise was welcomed at Caltech, Woolley expressed unabashed hostility toward the new radio techniques. When asked "Where do you think radio astronomy will be in ten years time?" at a talk on "The Future of Astronomy," Woolley replied simply "It will be forgotten." And after John Bolton published his list of six cosmic noise sources in *Nature* two years later, Woolley still dismissively remarked "Even if these objects did exist they could be of no possible interest to astronomers."[20]

Woolley's vocal distaste for the work of Bowen's Radiophysics Laboratory served perhaps to defend what he regarded as the institutional heart of astronomy in Australia. Despite his public aversion to the subject, Mount Stromlo hosted colloquia on extra-terrestrial radio noise, and even as Woolley was noted for his absence at them, he personally published on the topic in 1947.[21] In short, Woolley recognized the rising star of radio techniques emanating from the prestigious wartime Radiophysics Laboratory.[22] Woolley pointedly described that lab's astronomical work as "radio physics," i.e., as associated with physics rather than astronomy. Mount Stromlo, presumably, could then assume work associated with the astronomical significance of radio findings, while the boffins at Radiophysics continued to fiddle with their valves and antennas.

In England meanwhile, optical astronomers began embracing the young radio astronomers. They flushed at the grandiose plans of Lovell at Manchester, and fluttered over the far smaller program at the Cavendish Laboratory under Martin Ryle. We need to turn to Ryle's small, dynamic group to further explore the emerging transition of the radio physicists to the radio astronomers. Of course, the Cambridge group, like Lovell's and Bowen's, sought advice and approval from optical astronomers. In 1950, for example, Ryle wrote to an astronomer at the Edinburgh Observatory because Cambridge's work had, he said, "reached a stage of transition, when research of the exploratory type is combined with the starting of routine observations," including daily charts of the sun and regular recordings of a large number of galactic sources.[23] Early attempts at identifying radio stars and establishing possible correlations with visible objects brought Ryle and his colleague Frances Graham-Smith into contact with D. H. Dewhirst at the Cambridge Observatories. And, like Lovell, Ryle established his astronomical credentials with the broader British astronomy community by publishing in the *Monthly Notices of the Royal Astronomical Society*.[24] Martin Ryle, Frances Graham-Smith, and Anthony Hewish of Cambridge appear to be latecomers to radio astronomy, but we know that they had been refining their radio techniques in almost self-imposed isolation since the end of the war. By 1950, they too required community. Only once we see Manchester alongside Sydney and Cambridge can we recognize that community both legitimized novel work and gained respectability with potential students.

As foundational myths have been told and retold to eager listeners—David Edge, Michael Mulkay, and later Woody Sullivan—radio astronomers have remembered optical astronomers as dismissive of, or at best patronizing toward, the efforts of the radio researchers.[25] No point in the history of the radio astronomy community provides a better example of why we must treat the recollections of "those who were there" with care. When we rely on archival sources, a very different picture emerges— a picture not of antagonism and competition, but of community with a substantial degree of cooperation and interchange.

The Cambridge group, which Edge and Mulkay saw as intensely competitive, was, in fact, at least as indebted to the process of community formation as any other group. In late 1951, for example, Cambridge's Francis Graham Smith approached the Mount Wilson astronomer Walter Baade with newer, more accurate positions of radio stars and asked him to attempt correlations with optically observed objects. Believing Smith's measurements to be accurate enough, Baade agreed. For his part, Baade was nearing the end of a significant re-valuation of the size of the Milky Way galaxy and of the distance of our nearest galactic neighbors.[26] While Smith claimed an "accuracy . . . considerably improved by new interferometric techniques," Baade's search program was to encompass an area "3 times the errors of the present coordinates."[27] The optical astronomer was evidently wary of the veracity of the radio practitioner's claims. The first Cambridge survey, designated 1C, identified 50 radio sources with a claimed accuracy of within five minutes of arc. Though later the program would devote more attention to specific identifications, initially the group remained more interested in the apparent distribution of the radio objects across the sky, particularly their relationship to the galactic plane. Ryle and Graham Smith claimed in the *Monthly Notices*, and the claim was then repeated in *Sky and Telescope*, that the objects showed no concentration in the plane, indicating "that they must be extragalactic or relatively nearby objects." Though the Crab Nebula had been identified, no other supernova remnant could be, which implied a chance coincidence: "The authors conclude that radio stars represent a hitherto unobserved type of stellar body whose visible radiation is very weak."[28] The Cambridge opinion was doubted almost before it hit the presses. As early as October 1951 Baade excitedly reported finding an "exceedingly interesting object" in Cassiopeia. Comparable to nothing except the Crab Nebula, the object came within 1.9 minutes arc in declination and agreed perfectly in right ascension. Baade didn't wish to yet claim the radio source identified until he had done the Hα line measurements but admitted the coincidence appeared "cautiously suggestive." After re-photographing the area with the 200-inch telescope, Baade observed that the nebulosity extended further south than previously thought and so the positions came into alignment.

Still, even a "perfect coincidence" would not have "impress[ed]" Baade if the object were not so totally "abnormal."[29] Thus "normal" versus "abnormal" continued to be a central criterion for Baade's identifications for several years. In 1954, for example, one position of a radio source coincided with NGC 2623, which didn't display any abnormal characteristics on 48-inch Schmidt plates. Baade was again forced to spend time at the 200-inch telescope. His plates revealed two galaxies in collision, a definite abnormality, and hence he noted that the object could be a "radio emitter."[30]

If the Crab Nebula source captured the attention of optical astronomers, Baade's Cassiopeia findings held it. Moreover, a curious dependence emerged between the radio physicists in Cambridge and the astronomers at Mount Wilson. The nearly continuous correspondence between the California-based observer and the Cambridge-based listener over the course of 1951 and 1952 brought the Cambridge radio astronomy group into contact with the most powerful optical telescopes and some of the world's best optical astronomers, as had happened to the Australians two years earlier. The radio astronomers—who, as Graham Smith readily admitted, were "not experts on telescopes"—flooded Baade with positional data and readily acceded to his critiques about establishing the extent of the radio sources. But in turn, Baade became increasingly reliant on the radio data for his optical claims.

Locally, Ryle and Graham Smith benefited from the astronomical expertise of Dewhirst at the Cambridge Observatory, but Baade had no parallel recourse to a resident radio astronomer.[31] Dewhirst could not provide the optical resources of Baade in California, however. Across an ocean, the two halves of the identification effort worked together, because each side had technical skill the other lacked or instrumental access the other needed. It was in those moments that the radio astronomy community displayed its strength. Baade, the astronomer, and Graham Smith, the radio physicist, could have competed to locate optical correlates from either radio or optical data. Instead, the two men, working in separate institutions in different countries, managed to cooperate, to exchange data, and to obtain a successful outcome. They overcame the boundaries of different disciplinary conceptions of acceptable error bars, and each conveyed his specialty to the other.

The lesson of this long anecdote about the power of cooperation and community to create new knowledge challenges recent notions that different traditions of science, theorizing, instrument making, experimenting, and engineering "coordinate" their work but maintain "separate identities and practices."[32] In the Cold War world of physics, replete with vast black areas of secrecy, military funding, and compartmental-ization, one can easily see how separate identities not only continued but flourished. In the case of radio astronomy, however, overlapping interests between radio physicists and astronomers, and active cooperation between professional groups and between mentors and students, belie strict localization of scientific practice. Indeed, the experi-ence of Baade and Graham Smith illustrates the process of creating homogenous communities of practitioners transcending disciplinary, national, and expertise boundaries.

In later years, both Edward Bowen and Joe Pawsey would fondly recall that no one at Australia's Radiophysics Laboratory had had any professional astronomical training when they started radio astronomy. The standing joke became that the sum total of astronomical knowledge of the entire laboratory was one undergraduate textbook. Likewise, in Cambridge, Martin Ryle would remember the days when the optical community didn't believe that people who were "basically physicists" could become "respectable" astronomers, recalling the embarrassment of "frequently made silly mis-takes because we had no serious instruction in the elements of astronomy."[33] Ryle's right-hand man, Graham Smith, similarly recalled how neither he nor Ryle "had any training in celestial coordinates," and how they spent hours working out basic ele-ments of spherical astronomy.[34] Hanbury Brown too wistfully recollected considering astronomy "easy to pick up enough as he went along." These stories, told years later, have disguised how carefully and deliberately the radio physicists crafted their work to appeal to astronomical audiences. We need to appreciate the extent to which the leaders of the various programs were also disciples learning nearly as much as their students about this new intellectual and social field of knowledge called radio astronomy.

Embedded within the University of Manchester, Jodrell Bank was an educational facility where everyone, including Robert Hanbury Brown, engaged in training new

PhDs. The educational culture of Jodrell Bank grew from the fact that Jodrell Bank was institutionally a part of the university's physics department, and indeed Lovell himself was a member of the physics faculty. In practice what that meant was that Jodrell Bank radio astronomers acquired the training of the radio physicists more than any traditional astronomical education. Manchester's physics department considered a new specialized course on "The Physics of Experimental Method" useful to budding radio astronomers because it contained a topic on "electronic technique" and included substantial portions of amplifier design, feedback, and special circuit design. The department revised its final-year honors course in physics to permit students to take some "modern physics to a high level." Manchester's physics students chose from five topics: Physics of the Nucleus; Atoms, Molecules, and Quanta; Physics of Solid State; Geophysics and Astrophysics; and Methods of Theoretical Physics. It is significant that neither of the two staff members actively engaged in astronomical research, Bernard Lovell and A. J. Clegg, taught Geophysics and Astrophysics. Instead, two distinguished visiting lecturers, B. C. Browne from Cambridge and Erwin Findley-Freundlich from St. Andrews, divided up the course. Browne taught Geophysics; Findley-Freundlich taught Astrophysics. Lovell remained the primary instructor in Physics of the Nucleus, though it is clear that he kept a close organizational eye on Browne and Findley-Freundlich. Geophysics and Astrophysics covered "cosmic rays" as well as the "Structure of the Ionosphere" and the "Physical Properties of the Ionosphere." The final topic of the course, "Solar and Galactic emissions on radio wavelengths," had become Jodrell Bank's central research area, yet at first no Jodrell Bank staff member taught it.[35]

In contrast, Australia's Radiophysics Laboratory possessed no formal mechanism for internally training people in any science. The lab's umbrella organization, the CSIRO, lured people from the local universities and from abroad; invariably the senior scientists had doctorates from British universities. Aside from that, Joseph Pawsey (himself a Cambridge PhD from 1936) was aware of how few students of "the right types" were emerging from Australian universities in the early 1950s. Applicants to the Radiophysics Laboratory rarely were interested in radio astronomy research and rarely knew anything about the techniques involved. Pawsey bemoaned "the relative lack of the 'research student' type [to] carry the torch into the future." Always supportive of the Radiophysics Laboratory's efforts in radio astronomy, the eminent nuclear physicist Mark Oliphant believed that one or two of his own graduate students might be encouraged to sign up. The suggestion elated Pawsey: "If you felt that the arrangement could be satisfactory I should be very keen indeed to see them working in conjunction with some of our people. It would be a first-rate stimulus to our group. This raises the difficulty that it is desirable that the students should have simultaneous contact with (a) you and your people (b) astronomers (Stromlo) and (c) scientific radio people (Radiophysics)."[36] Pawsey saw the radio astronomer as a new type of student, one who existed in at least three worlds: the world of nuclear physics, the world of astronomers, and the world of radio physics. To become a radio astronomer one had to move easily among these institutions and disciplines, and presumably one had to face the Commonwealth Astronomer's hostility.

Communities establish their own patterns of discipleship. In order to fulfill the multiple roles expected of radio astronomy's disciples, the Radiophysics Laboratory had to send its practitioners abroad for training that combined astronomical knowl-edge with radio technique. An early example was Eric Hill, who applied to the Radiophysics Laboratory after graduating from the University of Melbourne with a degree in physics in 1951. Bowen arranged for a studentship so that Hill could study in Leiden under Jan Oort, then one of the world's premier astronomers with a deep interest in the new radio techniques. Significantly, Hill's major weakness, from Bowen's point of view, was that he was "not a radio man" and we might expect that Hill's "training" in radio astronomy would therefore have emphasized the radio-technical aspects of the science. Bowen, however, envisioned that after two years at Leiden Hill would be "exceptionally well endowed" to work in radio astronomy, since he would be the only "trained astronomer" in the entire group.[37] Pawsey hoped that Hill might "learn something of the spirit of the approach to research which [Oort and his] people demonstrate so effectively."[38] Possessing no background in radio electronics, Hill would unusual within the Australian group. Exactly what form Hill's training would take Bowen wouldn't venture to say. At the outset, Bowen hoped, Hill would "attempt to acquire a general background in astronomy." Expectations ran high as Hill, in the first of what Pawsey hoped would become a regular traffic in student exchanges, was sent in search of a style of research that the astronomically inexperi-enced members of Australia's Radiophysics Laboratory lacked.[39]

Another young radio physicist, Norman Christiansen, had been groomed to fill an intermediary role in the radio astronomy community. Seeking to educate the young radio physicist in the ways of astronomy, Pawsey offered to exchange Christiansen for a Dutch astronomer, perhaps Henk van de Hulst. Oort certainly thought that Chris-tiansen's visit would be worthwhile, even though there was little in the way of instru-ments for him, but doubted if anyone could make the return trip.[40] Thus formal personnel exchanges were another mechanism by which the Australian laboratory might recruit people. Indeed, by the mid 1950s, the Australian CSIRO, in collabora-tion with the British DSIR, offered postgraduate studentships for "training in research" in Australia. The amount of funds was roughly equivalent to British awards at the time; the Australian government sweetened the deal by offering passage on a tourist cruise ship for any award winner.[41]

In parallel to the Australian and Manchester cases, student recruitment also emerged as a major concern at the Cavendish Laboratory. That famous laboratory housed the Cambridge radio astronomy program, which attracted no students from the 1950 Cambridge Tripos II exams and only one from outside the university. Martin Ryle had arranged for all the Tripos Part II candidates to tour the field labs and the laboratory at Madingley Road, but even that attracted only three individuals. "Rather disappointing," said Ryle in a classic case of English understatement.[42] If the Cavendish couldn't attract Cambridge students, perhaps, Ryle hoped, they could pursue the Australian course and recruit from outside. Fearing that the Cambridge program would falter without additional researchers, Ryle wrote to Lovell at Manchester to ask, "privately," whether Lovell "[thought] it likely that any of [his] Finals students might be interested" in coming to Cambridge. Even in the austere Britain of the 1950s, lack of funding wasn't choking recruitment. A student should be able to secure a DSIR grant, Ryle said, if the student had "some radio experience (aerials rather than circuits)" and "an interest in astronomy."[43]

Ryle also reached further afield by contacting several students at Bristol University whom he believed would be interested. Importing students from outside Cambridge into the Cavendish presented its own problems. As he noted to Jack Ratcliffe, Ryle anticipated some "opposition . . . (e.g. from the Nuclear people) if we proposed a Cambridge D.S.I.R. grant for a Bristol man."[44] In the end, only one Bristol student accepted the offer, and he didn't show up. Students in Britain, like their cousins in the United States, flocked to nuclear physics for graduate work, and it appeared that the physicists would defend their new status to the last man and the last pound.

Competition for students became so fierce at Cambridge that several programs had to openly campaign to attract students merely to ensure continuation of research programs. In the late 1940s and the early 1950s, the Biophysics Unit at the Cavendish Laboratory, for example, found that "it was quite difficult to recruit good students." Like the radio astronomers, the biophysicists deemed it "necessary to 'sell' the Unit to good Part II students. [The] highest prestige area was nuclear physics, followed by low temperature physics in the early days, and then, at a later date, by radio astronomy."[45] While the bacteriophage group's leaders set up summer schools to attract new entrants,[46] Ryle's radio astronomy group received students who barely knew anything of the subject. One graduate student "wanted to do research in nuclear physics or low temperature physics," but when the Cavendish didn't have room for him he applied to Birmingham. Before he left, however, his advisor recommended that he meet with Ryle and consider radio astronomy. That student, John Baldwin, became Ryle's first graduate student in 1951.[47] Ryle remained "rather gloomy" about his prospects of being able to continue any substantial observing program at Cambridge, even after, at the last minute, several potential students appeared (probably because they hadn't been accepted anywhere else).[48]

The term "radio astronomy" itself, coined around 1949, conveys the joining of two of the scientific disciplines looking to uncover a vast array of new knowledge emanating at radio wavelengths from the stars and galaxies. As we have already gleaned, radio astronomy, in character and in culture, was far more complex than a simple meeting of the minds. People and institutions cooperated to create this new field. Optical astronomy lent its preexisting intellectual and social community, as Kopal supplied Hanbury Brown at Manchester with an education in astronomy. Meanwhile, radio physics donated technical competence and new vision via new instruments, as Hanbury Brown lent Kopal his positions for radio sources. Likewise, Bernard Lovell (Manchester), Joe Pawsey (Sydney), and Martin Ryle (Cambridge) presented radio physics techniques like astronomical instruments and practices, which astronomers, completing the circle, embraced more and more enthusiastically. Finally, all sought new disciples who might be trained in the new radio vision under a single sky.

Astronomers and radio physicists also participated in official international efforts to nurture the new social and intellectual community of radio astronomy. Both the International Astronomical Union and the International Union of Radio Science (officially the Union Radio-Scientifique Internationale, abbreviated URSI) contributed significantly to the new radio astronomy community. We might imagine a pitched disciplinary battle between the groups to secure the growing fortunes of radio astronomy for either radio physics or astronomy, but in practice almost every character we have met so far usually frequented both societies. The URSI's Commission V (Extra Terrestrial Radio Noise) burst onto the scene at the 1950 Zurich meeting. Six sessions of papers, talks, and heated debate saw the commission on Extra Terrestrial Radio Noise emerge as among the most vocal of the URSI's commissions. The meeting attracted a large cadre of international visitors, including Lloyd Berkner, Bernard Lovell, Joe Pawsey, John Bolton, Ken Westfold, Donald Menzel, and Edward Appleton. Donald Menzel from the Harvard College Observatory stands out as the only recognizable astronomer to attend the meeting. The Australian contribution was particularly strong, with Pawsey, Bolton, and Westfold all presenting, but work and comments flowed in from six member countries.

For many of the URSI participants, the topic of most interest revolved around radio propagation through the ionosphere and various radio techniques for detecting cosmic or solar phenomenon. The commission discussed radio wave scattering in the ionosphere and meteors before either solar or cosmic noise. From the perspective of the URSI, it seems that much of solar and cosmic noise had begun to fall outside that organization's purview. One of the first items of business, in fact, had been to "re-name Commission V 'Radio Astronomy'" but with the proviso that the area "extend its scope to include meteoric phenomena etc." In other words, Lovell's meteor results looked more like traditional research on radio propagation to the URSI Assembly than new investigations into either solar or cosmic noise. Although the URSI was primarily a professional body, its expansion to include radio astronomy paved the way to imagine a larger scientific community above profession, discipline, or nation. Indeed, the URSI and the International Astronomical Union entered into a "joint arrangement . . . for the world wide collection of solar data" and for "world wide radio observation of the sun." Both the IAU and the URSI legitimized the independent groups' work to an international audience. They also unified an expanding international network of researchers from a half dozen nations arrayed loosely under the headings "radio science" and "astronomy."[49]

Lovell and Hanbury Brown at Manchester reached out to the British astronomy community and made radio physics visible to an astronomical audience. The rest of this chapter inverts that story by moving the scene of the action to, first, one of the oldest American observatories, the Harvard College Observatory, and then, secondly, to one of the newest observatories, Caltech's. In both cases we see the opposite process of the first half of this chapter, the embrace of the radio physicists by optical astronomers.

The Harvard College Observatory, beset by declining visual conditions, became the first observatory to take up the idea, the technology, and the practice of the radio physicists and make radio a fundamental part of astronomy. Until Harvard's decision to build a radio telescope, the history of the radio astronomy community primarily concerned the radio physicists at Manchester, Sydney, and Cambridge taking their work to the astronomers and crafting radio physics knowledge into astronomical practices, instruments, and data. At the Harvard Observatory, and shortly afterward at Caltech, the process worked the other way around. There the optical astronomers employed a radio physicist to build a radio telescope and set about training young astronomers to operate the equipment while the astronomers trained themselves and their disciples in the proper place of radio within astronomy. Thus, by the end of this chapter it should be apparent that it took both trajectories—that of radio physics toward astronomy *and* that of astronomy toward radio physics—to create radio astronomy and to allow giant radio telescopes and the community to emerge and flourish.

It was, quite simply, the ability of radio to make visible the hydrogen spread throughout the universe that justified Harvard's investment in radio astronomy. Of all the possibilities that radio offered astronomy, few had as much allure as that of detecting hydrogen. As early as 1944, in the midst of the "hungry winter" in German-occupied Holland, Jan Oort and Hendrik van de Hulst at the Leiden Observatory concluded that the 21-centimeter hyperfine line of hydrogen should be theoretically detectable by radio. A quantum-mechanical phenomenon, the hydrogen hyperfine transition emanates from the interaction between the internal magnetic field produced by the motion of the electron and a spin magnetic dipole moment of the hydrogen nucleus. Though the Dutch speculated on the line's existence, they had had little opportunity to carry a program forward. Some abortive attempts in the few years immediately after the war with some old German radar sets had come to nothing. Later a small fire in a laboratory largely ended their initial research efforts. Moreover, the Dutch possessed little tacit radio expertise. Only by bringing in radio expertise in late 1948 to work with newer American radar equipment were the Dutch able to detect radio waves from the sun and then from the galaxy.[50]

To outside observers, the Dutch seemed well placed to detect hydrogen at 21 centimeters. As early as 1946, Greenstein threw cold water on Grote Reber's plans to discover the 21-centimeter emission line. Major work had already begun on the line, he said, Henk van de Hulst at Leiden having already "built a device" (though admittedly the Dutch had not "succeeded as yet" in detecting anything).[51] Greenstein, apparently convinced that van de Hulst's group would succeed in the first detection of hydrogen at 21 centimeters, was as surprised as anyone to learn that the Harvard physicists Harold Ewen and Edward Purcell had beaten the Dutch to the line in 1951.

Intellectually, the 21-centimeter line enabled astronomers to trace hydrogen through the heavens, opening an entirely new vista for the study of galactic structure. Modern astronomy now suddenly required new radio and electronic skills to equip observatories with radio telescopes capable of charting the line in the sky. Socially, it was a momentous result readily recognized by American and European astronomers. The result came from a pair of radio physicists who beat a long, detailed, theoretically driven search by astronomers at a notable Old World institution. Overnight, the line gave impetus to all-wave astronomy and forced American and European astronomers to cooperate fully with Australian and English radio physicists to realize the line's full potential to reveal the heavens.

While the Australians and the British cooperated with American optical astronomers and strove to detect radio sources and then correlate them with optically observed objects, the 21-centimeter line exploded the comfortable confines of the optical community. Jesse Greenstein at Caltech declared the observation of hydrogen at radio wavelengths "of extreme significance in astronomy."[52] Optical astronomers, especially those interested in galactic structure, responded enthusiastically to a new ability to "observe hydrogen gas in the vast regions of space where it is neutral." "The distribution of this matter"—hydrogen being the most common element in the universe—should "give clues to the evolution of our galactic system," *Sky and Telescope* predicted.[53] Reviewing the 85th meeting of the American Astronomical Society, held in Washington in 1951, *Sky and Telescope* gave more space to the 21-centimeter detection than to any other topic, and lauded the international nature of the new discovery. *Sky and Telescope*'s editorial discussed the Dutch work before revealing that the Harvard physicists Harold Ewen and Edward Purcell had recently detected the radiation. Yet the techniques and equipment they had used (especially their "horn antennas" and "superheterodyne" receivers) must have seemed strange astronomical equipment indeed to the Harvard Observatory's director, Harlow Shapley, and to his fellow astronomer Bart Bok.[54]

If we believe Ewen's and Purcell's memories of the 21-centimeter discovery, we must acknowledge that one of the most significant events of twentieth-century astronomy took place completely outside any astronomical institution and without the knowledge of any working astronomer. Yet it is hard to imagine that neither Ewen nor Purcell was aware that Henk van de Hulst was a visiting fellow at Harvard when the experimental runs took place, especially as Ewen and Purcell both acknowledged theoretical work he had done in the mid 1940s. Ewen and Purcell's experimental set-up consisted of a radiometer designed by Robert Dicke (another graduate of the wartime Radiation Laboratory at MIT) and two sets of crystal diodes, one from MIT's Research Laboratory of Electronics and one of much higher quality from Bell Labs. Ewen had bought, borrowed, or scavenged the remainder of the equipment from Harvard's physics department, especially the Nuclear Physics Laboratory. Moreover, Harold Ewen was a PhD student *in physics*, and spent most of early 1951 coaxing a coherent proton beam from Harvard's cyclotron.[55] He was, in short, even more of a physicist than the Australian or British radio physicists we have already met.

Ewen's horn-type antenna,[56] though it didn't have the utility of a steerable parabolic antenna, was cheap and effective. Its design restricted the wavelength of the received signal. Using the Earth's rotation, the horn would scan the sky, and Ewen would angle it to pass over the center of the galaxy. Unlike Hanbury Brown, who diligently worked to get more directionality out of his transit telescope, Ewen fixedly directed his gaze toward the center of the galaxy because he assumed it would have the most hydrogen and thus generate the most signal. He was not interested in any structure a hydrogen distribution might reveal, as Harvard's own galactic astronomers undoubtedly were, but only in the ability of his ultra-sensitive radio receiver to detect a faint radio frequency among the chatter. As historians of technology well know, similar technologies are often understood and used in radically different ways. Consequently, their users see radically different things.

Later in life, Ewen recalled resigning himself to writing up a negative thesis detailing the limit of his receiver's sensitivity.[57] Instead, through skill and hard work, Ewen succeeded on the evening of March 23, 1951, and the methods and implications of 21-centimeter radio astronomy ushered in, seemingly, a new age of vision. As Ewen concluded on his dissertation's first page, "detection of the hyperfine line is of considerable value to astronomical research since it provides information concerning the structure of our galactic system in regions not easily available by other experimental methods. A thorough research program will provide information concerning the spatial distribution and temperature of hydrogen clouds as well as a measure of their turbulent characteristics."[58]

To progress from a speculation in a thesis to an active research program would require, however, the cooperation of the radio physicists and the astronomers. When Ewen came to Purcell with the initial print showing the "telltale bump" of the 21-centimeter radiation, Purcell suggested that they seek outside confirmation. In other worlds, community was necessary to verify and develop the discovery. Ewen and Purcell contacted Henk van de Hulst (whom they seemingly now were able to locate) and Frank Kerr, an Australian radio physicist who was visiting Harvard. Both of those men wired home and got people to work to verify the results.[59] The Australians at the Radiophysics Laboratory had known of van de Hulst's theoretical predictions but were occupied by ever-finer source surveys; indeed, it appears that the detection took them by surprise.

Internationally, the discovery caused a sensation; locally, the experiment itself provided a springboard for Harvard's own radio astronomy project. Yet till that moment the power relationship between radio physics and astronomy had been heavily tipped in astronomy's favor. Lovell, Pawsey, and Ryle, as we have seen, had all consciously crafted their results and practices to emulate the astronomers. After 1951, however, traditional observatories such as those at Harvard and (later) Caltech launched major research and pedagogical programs in radio astronomy and actively sought out radio physicists. After the work mentioned in the preceding paragraph, Edward Purcell went back to Harvard's physics department; later he would receive a Nobel Prize for work on nuclear magnetic resonance. But Harold Ewen accepted a half-time appointment at the Harvard Observatory. It might come as no surprise that he chose a half-time appointment over continuing cyclotron work at Harvard, moving permanently to Los Alamos to work on the resurgent hydrogen bomb project, or dropping out of science and enrolling in law school. As I have emphasized, the culture of astronomy, and the new community of the radio astronomers, appealed to some radio physicists. The international culture of astronomy meshed well with the wartime expectations of the radio physicists, and together they would learn the shape of our galaxy.[60]

In the early 1950s, with radio astronomy in the vanguard, the once declining Harvard College Observatory shot to a position of prominence. Established in the nineteenth century, the Harvard College Observatory (HCO) forged its early culture from small numbers of students trained in intimate surroundings under close direction of their mentors while wealthy benefactors contributed buildings and telescopes. The institution steadily built a reputation as a center of astronomical research, not least because it employed a cadre of women astronomers as calculators and researchers. George Ellery Hale, who would go on to found the observatory at Mount Wilson, served as a voluntary assistant.[61] In the first half of the twentieth century, under pressure from newer and larger West Coast telescopes, particularly Hale's 100-inch at Mount Wilson, Harlow Shapley, the HCO's director from 1921 to 1952, succeeded in getting astronomy added to the list of subjects in which a PhD might be taken in 1922, and in 1929 Harvard began to offer formal PhD courses in astronomy.[62] As larger and better instruments came to dominate astronomical research, the Harvard College Observatory diversified its research interests into meteors, galactic structure, and stellar structure. By 1939, Harvard was one of only fourteen American institutions at which astronomers were trained; by 1958, Harvard graduates accounted for one-fourth of the astronomy PhDs produced in the United States.[63]

Long aware of the declining condition of the local night sky from light and industrial pollution, Harvard's astronomers had engaged in numerous cooperative ventures to gain access to better instruments and better seeing conditions. Fred Whipple cultivated connections with the military, but his colleagues Harlow Shapley, Bart Bok, and Cecilia Payne-Gaposchkin remained extremely reluctant to develop further military patronage ties, seemingly for political reasons.[64] If there was personal tension within the observatory family, much of it may have been due to the faltering of the HCO's connections to other facilities. A long-standing partnership with an observing station in South Africa collapsed in 1953 when Harvard withdrew financial support over the issue of Apartheid. Donald Menzel briefly took part in a joint venture in solar astronomy with the Air Force at Sacramento Peak in New Mexico, but after Harvard refused to appoint a full-time staff member to the observing station and the arrangement fell apart; the Air Force continued the work without Harvard.[65] This pattern, the historian David DeVorkin tells us, became familiar throughout postwar astronomy; indeed, many astronomers began to talk not of having new instruments for individual institutions but of building cooperative facilities underwritten by the federal government.[66]

By the early 1950s, graduate students were the dominant users of Harvard's local telescopes. The "principal significance" of the HCO's four minor telescopes (the 20-inch reflector, the 16-inch Metcalf, the 12-inch Metcalf, and the 8-inch Ross) was as "training instruments" on which, Bart Bok claimed, "our students must cut their teeth."[67] The Boston area's deteriorating optical conditions and the need for adjustments to Harvard's major local research telescope, the Jewett-Schmidt, were detrimental to the entire observatory. Even after a substantial refurbishment program, by 1952 Bok found most of the optical telescopes at Harvard's Agassiz Station (located in the town of Harvard, Massachusetts) wanting. The 61-inch Jewett-Schmidt telescope still had "an inadequate slit in the roof, which does not permit observation near the zenith, a too-short polar axis, which makes the skies around the celestial pole almost inaccessible, and a poorly-engineered mechanism for guiding in declination." And this was the instrument, Bok affirmed, that had "done good work in the field of galactic and extra-galactic research and is continuing to do so." In the late 1940s, Bok didn't pursue any major readjustments to the telescope's mounting, preferring to "postpone such changes to the time of the transfer of the telescope from its present location to a more favorable site."[68] In short, the poor local night sky and the telescope's general disrepair were major impediments to any local research program. Bok awaited the day when the telescope could be relocated, but the dominance of newer West Coast instruments made it unlikely that any research justification would ever make it worthwhile to move the telescope.[69] In short, one can only agree with Jesse Greenstein that by the postwar years "the generation of high precision data was not a Harvard specialty."[70]

Not surprisingly, the declining status of Harvard's observatory affected its internal ability to maintain staffing levels and to obtain funding. As early as 1948, Bok complained that "Dr. Shapley's galaxy bureau [had] been reduced from eight assistants to one and one-third assistants." His own "Milky Way research" had been "reduced from four to a fraction of the time of a single secretary-scientific assistant." Similar reductions occurred in the variable star bureau. Stellar photometry was "abandoned."[71] In 1951, approaching retirement, Shapley had to resort to teaching in the summer. It was, as Fred Whipple mentioned to Greenstein, "a new departure of the first order," indicating both the significance of teaching and the sad state of astronomy at Harvard.[72] Whipple had misgivings about letting Shapley teach, but both Shapley and his students seemed to enjoy the experience.

The new importance of student recruitment and training, and the Harvard astronomers' new role as teachers, were indicative of significant shifts in the culture of American astronomy. Early in 1952, the HCO's new director, Donald Menzel, characterized Harvard's local observing site, the Agassiz Station, as follows: "[It] is used for a number of researches, both by major staff and by graduate students. It is, moreover, an important item in the training of graduate students, since most colleges or universities hiring a new PhD will be interested in a man who has some observational as well as theoretical training. This station enables us to carry on our graduate school effectively."[73] By the early 1950s, Harvard's own faculty recognized that their observatory's "relative standing" was being "challenged by our own former students." Harvard graduates were now "important members" of the "four other active graduate schools of astronomy in America."[74] Of course, it is not surprising that Menzel—a traditional optical astronomer—continued to emphasize the research output of a station that obviously had been eclipsed.

In 1952, Bernard Lovell secured the funding for his new 250-foot "radio telescope," less than a year after Ewen and Purcell made interstellar hydrogen visible at radio wavelengths. As Harvard's sky became opaque, an entire new class of astronomical instruments that could see through the pollution became available to astronomers. That year, Harlow Shapley, speaking as a member of a passing generation, outlined two possibilities that he thought would restore the Harvard College Observatory's standing: "a super-objective prism radial velocity telescope, or an adventure into radio astronomy." Shapley himself dreamed of "a southern reflector of diameter 100-inches or greater." He longed to realize his "larger instrumental dream,"[75] but "the [Observatory] Council shied away from the suggestion . . . largely because of the almost certain ten to fifteen years' waiting period." "We are too old," Shapley lamented.

We know from the minutes of their meetings that the members of the Observatory Council wanted to pursue the objective-prism radial velocity project over any 21-centimeter research. But dreams of new optical instruments remained fanciful in the local context of poor skies and poorer budgets. Instead, the decision from the Observatory Council, though "not because of lack of interest," was to "postpone" radial velocity work and to support radio astronomy in the United States.[76]

This was not a competition between research programs—either a radial velocity telescope or a radio telescope could be used for galactic research. Rather, it was a question of technological and pedagogical cooperation. Locality permitted radio astronomy to triumph over its radial velocity rival at Harvard because radial velocity work necessitated ownership of, or at least access to, a large and expensive new telescope somewhere else, probably in the southern hemisphere because seeing conditions around Boston were so poor. Radio, of course, did "not require excellent climatic conditions"[77]—precisely the argument Bernard Lovell used to sway the British astronomy community. Moreover, once the definition of 'telescope' had changed to include radio instruments, a new telescope could be built at or near Harvard. A new telescope would preserve a long-standing research program dedicated to studying the physical characteristics of the galaxy, particularly the distribution of hydrogen, and would not radically alter the priorities of Harvard's astronomers. Indeed, Shapley, Bok, and Menzel foresaw vast benefits for their institution. "Radio astronomy could grow here as optical astronomy has done in the west," Menzel fantasized.[78] The clear skies of California had helped the 100-inch Mount Wilson telescope dominate astronomy for a generation, and the 200-inch promised to ensure West Coast superiority; perhaps, considered as 'telescopes,' the planned 24-foot dish and a future 60-foot dish would do the same for East Coast radio astronomy.

Harvard's radio astronomy program took shape rapidly after Harold Ewen moved to the HCO. In 1952, shortly before retiring as director, Harlow Shapley noted the "exceptional progress in the development of equipment," especially "the installation of a 25-foot mirror (dish) for researchers in radio astronomy." The physicist Ewen had constructed the radio receiver, the "black box," that formed "the heart of this new-type astronomical research instrument," Shapley exalted. Radio astronomy at Harvard began with Ewen's "magical electronic recording devices."[79] Novel and unique instruments served as advertisements for a strong scientific program. Harvard University's president, James Bryant Conant, lauded the Agassiz Station as "the only astronomical observatory in America equipped with a radio telescope at this time."[80]

Shapley's rhetoric—emphasizing the "astronomical" nature of the "equipment"—belied a deeper truth. The technological choice between new instruments fundamentally changed the culture of astronomy. Bigger instruments and more physics became necessary as lures to draw new students. Students became the moral capital of astronomy, not merely components of a system of knowledge production. As Bart Bok impressed upon Harvard's Observatory Council, a stronger commitment to larger antennas and greater physics content would be necessary if the HCO wished to maintain its "appeal to graduate students in radio astronomy."[81] Bok and Harvard tied the social capital of astronomy to the recruitment and training of students. Moreover, Bok found an important patron interested in supporting facilities for graduate students, the National Science Foundation. As a method of democratically contributing to needs of the nation the NSF sought to expand educational opportunities in science. Bart Bok believed that the NSF was "very much interested in promoting graduate education in astronomy and might well be receptive to an appeal that they help pay for putting [Harvard's] Agassiz Station in tip-top condition as a center for graduate student training."[82] Thus their adventure into radio astronomy rapidly fulfilled its promised by giving the HCO a new instrument to draw students, and with students the potential for new sources of patronage.

Though astronomers viewed the coffers of the Office of Naval Research and the National Science Foundation more cautiously than other scientists did, especially the physicists,[83] money for graduate instruction was, for the astronomers, a morally unproblematic way of advantageously accepting federal largess. Of course, the astronomers would have to alter the meaning of 'astronomy' so as to treat optical and radio telescopes as equivalent. As Bok noted, to strengthen observational astronomy "the training of astronomers" had to be "not only in the operations of telescopes but in making the scientists familiar with the newest techniques of physics and engineering that may have a useful application for astronomy."[84] To reassert astronomy, the new disciples of astronomy would become far more interdisciplinarily adept than their forebears. Soon afterward, no doubt anxious to reassert the observatory's status and to "attract some converts,"[85] Bok decided to hold an "informational and inspirational" symposium on radio astronomy at the 1953 annual meeting of the American Association for the Advancement of Science. The symposium was only a limited success, however, as Bok's drawing card, the Cambridge University radio physicist Martin Ryle, was unable to attend. Still, one of the oldest observatories in the United States inviting a man with no formal astronomical training as a star speaker demonstrates how much the culture of astronomy had already changed.

A patron's gift in early 1954, Bok said, put "the 21 centimeter Radio Astronomy Project at Agassiz Station a solid footing." Indeed, the building of Harvard's first radio telescope opened "the door to much wider development in Radio Astronomy at Harvard." The wider development brought new opportunities for research and strengthened Harvard's recruitment and training of disciples of astronomy. Together, said Bok, research and pedagogy would "strengthen terrifically our East Coast claim."[86]

In fact, the disciples came before the research. Even before any radio telescope existed, radio astronomy became a part of the astronomy curriculum at Harvard. This point requires added emphasis: Pedagogy exists in the vanguard of scientific research, not the rear guard. Earlier histories have claimed, like Ronald Doel, that no "curriculum change" took place in astronomy (in solar-system astronomy, anyway) "through the mid-1950s," so that the impact of radio astronomy on the field in the United States was effectively postponed until the "late 1950s."[87] Yet the first course in radio astronomy at Harvard—one of only four American university granting PhDs in astronomy—predated the completion of the 24-foot radio telescope at the Agassiz Station.

Changes in the curriculum immediately brought about a changes in the astronomy community: Because no Harvard astronomer was qualified to teach Harvard's new course in radio astronomy, the job of teaching the course was given to the visiting Dutch astronomer Henk van de Hulst and the Australian radio physicist Frank Kerr (himself ostensibly learning astronomy). Their initial course, which attracted five graduate students, aimed to introduce physicists and astronomers to "the problems and possibilities" of radio astronomy. The students wrote up the final versions of the lectures as their assignments, and two of van de Hulst's graduate students back in Leiden edited them for publication.[88] Not merely passive recipients of established knowledge, the students helped construct instruments, lectures, and even an early textbook on radio astronomy alongside their teachers, fuzzily remaking new knowledge and social arrangements in astronomy. Here we instantly recognize an extension of Andrew Warwick's insight that "examination problems served to generate community."[89] In radio astronomy, the act of writing the lectures laid a foundation for an international, interdisciplinary and intergenerational community. When van de Hulst and Kerr left the following summer, Bart Bok took over the courses on radio astronomy and built a new science. Each year, as part of a "general survey of the whole field of Galactic Radio Astronomy," Bok gave two supplementary lectures on "Basic Galactic Structures." But when the course changed to include radio techniques, particularly 21-centimeter-line frequencies, Harold Ewen (a co-discoverer of the line) stepped in.[90] Thus, even a master of galactic astronomy became a disciple of radio astronomy in those early years.

Grad students' active participation in the construction of their own courses served as a prelude to the first instruments of radio astronomy at Harvard. Harvard's astronomers overtly encouraged "physicists and astronomers" as well as "advanced astronomical students" to take "part in the planning, building, and operation of the new tool" of astronomy.[91] Construction of the new radio tools went forward in concert with courses involving new radio electronics as well as spectroscopes and photographic plates. All served to reinforce Harvard's established pedagogical philosophy of familiarizing students with the myriad instrumental possibilities in astronomy. Once built, the new Harvard radio telescope remained "in the hands of three advanced students, two of them gathering material for doctoral theses in the field of 21 cm cosmic hydrogen radiation." In December of 1952, David Heeschen, one of these new disciples, represented Harvard astronomy to the American Astronomical Society with a paper titled "Calculated Line Profiles for the 21-cm Line of Hydrogen."[92] Though such a title may sound more akin to a physics presentation on the properties of the line profiles, the work of Harvard's students, like that of earlier students at Manchester, was firmly rooted in traditional cosmic concerns: Bok told the "Radio Astronomy Men" of Heeschen's "scientific results," which "relat[ed] to the ratio of gas and dust in interstellar space and to the physics of the interstellar gas."[93]

The new radio telescope and the courses in radio astronomy quickly "aroused a good deal of interest among our graduate and undergraduate students"[94]—and not only interest, but success too. After only two years in operation, Harvard's radio astronomy project began producing radio astronomers—a fact that Bok and Ewen highlighted when applying to the NSF for funding with which to expand their radio telescope. "Several students at Harvard and elsewhere have expressed an interest in research in Radio Astronomy and we expect to have a steady flow of PhD candidates in the field. Since we continue to look upon the Agassiz Station Project as serving the dual purpose of research <u>and</u> training in Radio Astronomy, we shall continue to encourage wide participation by qualified graduate students."[95] Late in 1954, David Heeschen and A. Edward Lilley became the first American PhDs in radio astronomy.[96]

The new character of radio astronomy at Harvard quickly altered the culture of astronomy. Of course, a new culture is often defended as a better version of the old one. At Harvard, radio quickly took a place alongside photography as a method of astronomy. In the popular magazine *Sky and Telescope*, Bart Bok wrote that "the radio technique has taken its place along with the photographic, photoelectric, and spectrographic methods."[97] For Bok, at least, radio astronomy remained a *technique* to answer the *questions of astronomy* and support the "graduate training program." The very success of Harvard's radio astronomy program had caused the university's administration to question the direction of the Observatory. Bok himself, as the de facto leader of the HCO's efforts in radio astronomy, found himself compelled to justify the new disciplinary arrangements:

I took the opportunity to make very clear to the Council that I am still primarily a Milky Way astronomer who happens to be concentrating <u>at the moment</u> on radio astronomical studies relating to galactic structure, but who feels that graduate study in galactic structure can only

FIGURE 3.1
The Harvard Observatory's first radio telescope, Agassiz Station, Harvard, Massachusetts, c. 1953. From B. J. Bok and H. I. Ewen, "Radio Astronomy in the Microwave Region," November 10, 1953, UAV 630.452.15, Harvard College Archives. Reproduced courtesy of Harvard College Library.

be effective if it offers the student a balanced training program, which should include wide-angle photographic work with and without objective prism and experience in photo-electric research along with radio astronomical work. I made it very clear that I consider radio astronomy a technique—one that incidentally is just as useful to the solar physicist, or the meteoric astronomer, as to the galactic astronomer.[98]

Within only two years, the Harvard College Observatory—a venerable, traditional institution—could consider the instruments and practices of radio astronomy as much a part of astronomy as the photographic plate. Likewise, they could view operating a radio telescope as an effective education for a budding young astronomer. Bok's defense of radio astronomy as simply a new technique of astronomy also preserved the burgeoning community of optical and radio scientists. As the radio physicists crafted their instruments and data to cooperate with the astronomers, so Bok the astronomer normalized those new techniques within astronomy.

We must remember that these changes came about because of demands for recruitment and training of students. Both optical and radio astronomers had to bow to the pressure of pedagogy. In a telling battle of boundaries, Bart Bok insisted in 1955 that the radio telescope and receivers remain in near continuous working order. Harold Ewen, on the other hand, wanted to test "new solar radio equipment," and "stop regular observation with our 21 cm equipment during December and January." The experiments could wait, Bok asserted, until the 60-foot radio telescope was in operation, because of the importance of the new endeavor to the graduate program. "If we were not to have the equipment in good 21 cm operating condition by about February, then I fear that one or two of our most promising candidates may shift to other fields."[99] Even the threat that such limited numbers of students might "shift to other fields" was enough to halt Ewen's plans for further instrument experimentation. Then again, in a small coup, Ewen eventually won (in 1956) a long argument over "the necessity of a greater knowledge of electronics on the part of graduate student." He even managed to introduce the requirement that Harvard's radio astronomy students take "one or more courses in electronics in the Applied Physics Department."[100]

In less than a decade, and before Sputnik would again change the culture of astronomy, one significant American observatory took an adventure into radio astronomy. In integrating radio into astronomy, Harvard's Observatory opened its doors to an international array of interdisciplinary scientists to reassert its position via the recruitment and training of new disciples. Radio electronics became a standard part of Harvard's curriculum and the radio telescope a standard instrument of astronomy. These changing notions of what "astronomy" meant at Harvard, in turn, redefined the idea of the radio astronomy community. Radio astronomy now was no longer merely an electronic addendum to traditional astronomy. Radio astronomers had altered the measure and meaning of astronomy itself. And Harvard now had a 60-foot radio telescope.

On the opposite coast to old Harvard, the modernist California Institute of Technology was, its most famous astronomer Edwin Hubble claimed, the "world's center for astronomy." Caltech had become a major research university in the interwar years through the efforts of two of science's greatest boosters, Robert Millikan and George Ellery Hale. Hale's own field of astrophysics was to be not merely a union of physics and astronomy, but a triumvirate of astronomy, chemistry, and physics.[101] Hale garnered a reputation for capturing the imaginations of wealthy philanthropic businessmen, particularly Andrew Carnegie. Hale, of course, is now remembered for having planned the next large telescope before the previous one had even been completed: the 100-inch as the 60-inch went up, then the 200-inch before even the foundation of the 100-inch was finished.[102] Each opened the visible sky to astronomy as never before.

Though Caltech's telescopes were the envy of old Harvard, Caltech's astronomers confronted exactly the same dilemma that Harvard's faced: by the early days of the Cold War, it had become all too apparent that astronomy was hampered by a dearth of disciples rather than an abundance. Traditional astronomy aroused little interest among the public, or even among college-bound students. According to Jesse Greenstein, Caltech, though in "the unique situation" of possessing "the three most powerful telescopes of the world,"[103] continued to struggle to recruit students. The world's largest, newest, and most expensive optical telescope sat atop Palomar Mountain, but its few triumphs remained overshadowed by all manner of new technology. With its Art Deco appearance, the 200-inch telescope appeared antiquated in comparison with the futurist radars, jets, and all things atomic so celebrated in comic books and popular culture. Only four years earlier, with the merger of the Caltech and Carnegie observatories and the completion of the 200-inch telescope, Ira Bowen's chief worry, as director of the Mount Wilson and Palomar observatories, was the impression that the observatories were deliberately poaching students from fellow programs. In 1952, however, Bowen lamented to Vannevar Bush the "failure to get men . . . into astronomy," at least partly "caused by the glamour of nuclear physics and electronics."[104]

Much of the history of Cold War–era science has been concerned with the glamorous science of nuclear physics and its own significant cultural changes brought on by its newfound popularity and prestige. David Kaiser, for instance, attributed the popularization of Richard Feynman's diagrams to the arrival of so many new faces in so many new departments that teaching by the traditional and much-romanticized method of personal mentorship no longer was practicable. Feynman, a member of Caltech's physics department, thus lent his name to a system that had been developed because the oversupply of students influenced the choice of available methodological tools available to future physicists through their education.[105] Paradoxically, the culture of physics suffered from its own success. There were simply too many new faces, and the character of physics was irrevocably altered as intimate, personalized instruction gave way to large, generic, anonymous lectures. Physicists became black boxes, trained so that the military-industrial complex could call on them at any time to solve the immediate problems of the day.[106]

The 1950s was the golden age of physics. There were two jobs for every physics PhD, and the world wined and dined physicists as dignitaries. The American accelerator authority M. Stanley Livingston once noted the "employment problems of young physicists": there were too many jobs. The records of the American Institute of Physics indicate that until the mid 1960s demand far outstripped supply of physicists in all sections of the military-industrial-academic complex.[107] With abundance, the culture of physics changed as new recruits dreamed of well-paying jobs in industry.[108] American intellectuals saw the rise of bureaucracy as seeming to signal the death of individualistic American pragmatism and created a mythical longing for the lost European world of close-knit groups of physicists in small cloistered towns such as Göttingen. Intellectuals aside, the majority of physics students voted with their feet: by 1957 "more than 85 percent of [American] undergraduate physics majors listed industrial jobs as their top choice."[109]

In contrast, the culture of radio astronomy was driven not by abundance but by poverty—especially the paucity of students. Lee DuBridge, Caltech's president, enthusiastically launched that institute's radio astronomy program in the middle of 1952, not two months after Ira Bowen's dire pronouncement to Vannevar Bush. Caltech's "proximity to large telescopes" made it an ideal location for radio astronomy, or so Australia's Edward Bowen had convinced DuBridge.[110] In the early 1950s, Ira Bowen, director of the Combined Observatories, cautioned against further fragmenting astronomy through the addition of radio astronomy at Caltech. Certainly astronomy had benefited from the marvelous new technological options that had become available after World War II. But all those wonderful technologies, exactly as Arthur C. Clarke had predicted, had caused a regrettable "temptation to attempt too many observational programs" with too few people.[111] For Ira Bowen, the simple existence of a new technology didn't immediately imply the adoption of any or all of them. In astronomy, the appearance of too many radically new techniques and instruments—radio astronomy, image tubes, rockets, electronic computers—threatened to yet further fragment the Combined Observatories. In contrast to every other observatory, the Combined Observatories suffered from an almost embarrassment of luxuries when they looked at their instrument armamentarium. Yet the mere possession of the largest optical telescope didn't automatically assure Caltech–Mount Wilson status or security. Effective use of time at the 200-inch Palomar Mountain telescope required a substantial visitors' program, which additionally bolstered the institution financially and intellectually.

In light of Ira Bowen's initial resistance to radio astronomy at Caltech and his later enthusiastic support, it seems reasonable to suppose that his acknowledgment of the failure to recruitment students must have played a large part in swaying his change of mind. Caltech's astronomers, then, seem to have speculated that radio astronomy appropriately attached to the world's major optical observatory, might attract new disciples, but only if effectively clothed in the garb of an international and interdisciplinary community. We know that the close personal relationship between Lee DuBridge and Edward Bowen seemed to smooth the way for a new era of cooperation between Australian radio physicists and the American astronomers. Edward Bowen actively supported the creation of a radio astronomy program at Caltech, seeing a Caltech program not as a competitor but as a necessary part of the radio astronomy community. Simultaneously, he convinced Vannevar Bush, chairman of the Carnegie Institution, the primary patron of the Palomar Mountain 200-inch telescope, that Australian and American radio astronomy would become the world's foremost "collaborative effort, in which the work in the southern hemisphere [would be] complementary to that in the northern."[112] This unification of southern and northern observing sites serves as a geographical metaphor for what was about to happen in radio astronomy: collaboration between north and south, between radio physicists and astronomers, and between Australia and the United States.

To the Caltech astronomer Jesse Greenstein, it was precisely those "radio engineers and radar experts, and recently young physicists" that astronomers ought to contact in order to reverse prospective new disciples' seeming lackluster interest in traditional astronomy.[113] Likewise, in the wake of the debacle over Edwin Hubble's failure to become director of the Combined Observatories, Ira Bowen emphasized to Bush that astronomy's future lay with "new concepts or new experimental techniques in Physics." Bowen saw one part of the work of the Combined Observatories as moving into "applications of nuclear physics to astronomical problems," particularly with regard to the abundance of elements in stars.[114] In the early 1950s, however, openly welcoming eager visitors presented a ready-made solution. In rapid succession, Caltech and Mount Wilson hosted Edward Bowen and Joe Pawsey from Australia and Frances Graham-Smith from Cambridge.

Like Bok at Harvard, Greenstein fought to have radio included in Caltech's definition of astronomy. For Greenstein, radio's substantial presence in the major current work of astronomy, especially the "correlation of point sources work and with galactic structure and astrophysics," demanded that optical astronomers cooperate with the radio physicists.[115] In other words, it was not that radio presented the sole option for the future development of astronomy; it was that even the astronomers at the world's foremost observatory looked toward new technologies and interdisciplinary arrangements as their science's future.

DuBridge evidently agreed, and he wasted no time. He demanded to know the scale of the commitment that Caltech would have to make in order to begin work in radio astronomy immediately. He quizzed his old Radiation Lab buddy Edward "Taffy" Bowen on the basic problems in radio astronomy. What "group of minimum instruments," he asked, would be "required for the initiation of the laboratory?" DuBridge never thought small. Imagine the future, he prodded. What "somewhat more ambitious plan for a major large instrument" could he foresee?[116] This rapid move from a minimal initial investment in equipment to conceptions of a grand project must be a key characteristic of the drive toward "big science" in the Cold War. There was a palatable fear of being left behind, and of students' being lured away. We should also note that Caltech actively participated in the making of the military-industrial complex, since under DuBridge's governance Caltech founded the Jet Propulsion Laboratory that eventually would take Americans to the moon. The change from small science to big science might be better characterized as the demise of the Cambridge tradition as Cold War institutions, like their national governments, expressed leadership through scientific and technological dominance. The social organization of science in the "community style," as opposed to the "complex style," didn't negate large instruments, teams of researchers, or big budgets. It actually expanded opportunities because communities didn't have to be hemmed in by artificial boundaries of nationalism, discipline, or secrecy.

For DuBridge, the future of radio astronomy at Caltech came down to one question: Would Caltech be building "a new laboratory or observatory of radio astronomy"? DuBridge zeroed in on the real possibility that radio astronomy could be a laboratory science rather than an observatory one. The idea itself gives us pause to ask what scientists in the Cold War era considered the difference between a "laboratory" and an "observatory" to be. Once again, the answer broadens our vision of science in that era, because the difference between a laboratory and an observatory specified the distinctions between the community style of science and the complex style.

Vannevar Bush advocated "an observatory rather than an instrument" for radio astronomy at Caltech, for example. An observatory, Bush said, retained a set of complementary instruments, techniques, and practitioners unified under a single research program. For Bush, the lack of a single evident instrument of radio astronomy differentiated a hypothetical radio astronomy laboratory from a radio astronomy observatory. As Bernard Lovell was committing hundreds of thousands of pounds to the Jodrell Bank radio telescope, Vannevar Bush doubted that in the immediate future any single form of instrument would present itself to radio astronomy. He envisioned long periods spent with several instrument choices "before one could concentrate with assurance on a relatively large and universal instrument."[117] Likewise, even after Caltech seemed to have come down firmly on the side of an "observatory," Robert Bacher, chairman of Caltech's Division of Astronomy, Physics, and Mathematics, still couldn't say just "what sort of a Radio Observatory would be desirable."[118]

Envisioning an "observatory" meant envisioning the "radio telescope." With Lovell, Manchester's radio astronomers had negotiated the meaning of their instrument such that radio data presented astronomical information in an astronomically recognizable way. In California, Vannevar Bush suggested that Caltech might erect a 200-foot radio telescope as an analogy to the 200-inch Palomar optical telescope. That pleased Ira Bowen.[119] Beyond the utility of an advertising gimmick, however, Bowen seriously questioned the form of the radio telescope as necessarily similar to the optical telescope. Just "because large circular mirrors are the standard light collectors for optical astronomy does not mean that they are necessarily the optimum type of receiver for radio astronomy." It was thus the astronomer Ira Bowen who advanced a significant insight into the emerging new measure and meaning of astronomical technology:

[If] correlations are to be developed between [radio] observations and observations in the optical range of wavelength . . . the accuracy with which a radio source can be located in the sky is directly proportional to the resolving power of the equipment available. The success of this equipment for problems of the type proposed will depend very largely on the attainment of sufficient accuracy in the location of various radio sources to positively identify them with optical observed objects. Failure to do this has been one of the major factors in holding up cooperative studies between radio and optical studies in the past.[120]

In short, at Caltech, optical astronomers moved beyond merely astronomically recognizable data to a demand for greater resolving power for radio astronomy so as to standardize the twin approaches of astronomy, optical and radio. With greater resolution, astronomers' ability to locate and correlate radio sources with optically visible objects would become the scientific focus of the entire astronomy community. In order to gain both increased resolution and successful identifications, Bowen advocated the construction of high-resolution equipment. Thus, cooperation between differing ideas of precision by radio and optical methods became the essence of the new social arrangement between international disciplinary practitioners. It is little wonder, then, that, from atop Palomar Mountain, Ira Bowen insisted on "the closest cooperation with other groups, particularly the one in Australia."[121] Reciprocally, Edward Bowen reminded Lee DuBridge on several occasions of the benefits of cooperation and exchange between the Australians and the Americans. Unlike those "at Manchester," Bowen noted, who "will not have the benefit of a close association with good visual observations or, for that matter, with other astronomers. And, due to the bad seeing conditions, they are unlikely to get it."[122]

So important was the new community combining optical and radio astronomers that Jesse Greenstein and Rudolph Minkowski feared that any entirely independent Caltech radio astronomy program would harm the relationship. They had witnessed, they said, the fragmentation of intellectual and social communities when sciences had begun to use new scientific instruments and to accept disciplinarily external practitioners. True, Greenstein envisioned long-term benefits emerging from a Caltech program in radio astronomy, but he questioned why it should be independent of the Australian and British groups. "As an astronomer," he wrote in notes for a now-unknown purpose, the Combined Observatories' relationships with radio astronomers were "essentially perfect." "We are in contact with all active groups, receive valuable data long before publication, and have cooperated with rival institutions successfully and impartially."[123] Likewise, Walter Baade, one of Mount Wilson's most senior astronomers, worried that "existing radio astronomy groups might not be quite so free to turn over material prior to publication and to discuss things freely if we were an active part of a competing group."[124] Radio astronomy had succeeded, Greenstein, Baade, and Minkowski appreciated, via a community experience that saw regular and free exchange between optical astronomers and radio physicists. In turn, the emphasis on cooperation between international and interdisciplinary groups provided the context for the shape of Caltech's radio astronomy program and actively informed its instrument choice, intellectual pursuits, and social organization. In short, competition didn't drive the production of scientific knowledge in radio astronomy; sharing of data did.

Once more, the demands of the new community affected substantive changes to graduate course material and content. Well aware of the appeal of electronics, astronomy's disciples were being taught radio astronomy by 1954. For example, a graduate course taught by Robert Bacher, the chairman of Caltech's division of Physics, Mathematics, and Astronomy, culminated in "a discussion of the point sources, identifications and theoretical interpretation." Because Bacher was an optical astronomer, his course covered ("somewhat non-technically") antennas, receivers, and "interferometers, et cetera," and outlined solar radio astronomy "from the observational point of view, together with sufficient background on solar optical phenomena to show the contributions possible in this field."[125] The content may not have been as important as Bacher's approach: Caltech's graduate students in astronomy would emerge from their training with an eye toward the visible, but the course presented only a limited view of the technical, electronic components. The evident pedagogical lesson emphasized cooperation and collaboration between the radio physicist and the astronomer to aid in the discovery and interpretation of unusual objects observed by radio astronomers. Leaving no room for a conception of competing traditions of optical and radio astronomy, Bacher's course secured all-wave astronomy in the minds of Caltech's disciples of astronomy.

In 1955, Jesse Greenstein revised the entire graduate curriculum extensively. Pedagogical changes marking the inclusion of electronics in Caltech's astronomy curriculum resulted in the appointment a new faculty member: Arthur D. Code. Code secured his employment in the Department of Astronomy by being "not only one of the most capable young astronomers in the country but . . . working in a field that is very essential to strengthen in both the Department of Astronomy and the Observatories and that is, the field of electronic techniques." Electronic techniques were, Ira Bowen believed, "rapidly increasing in importance in astronomy and it is very essential that the Institute be in a position to provide the coming young astronomers with training in the field."[126] Code's course, "Astronomical Radiation Measurements," added a full-year formal course to the graduate program, pushing the number of formal courses to three. The course concentrated on all manner of astronomical detectors, from photoelectric to emulsions to radiation detectors, and in the second term students were required to build their own amplifiers.[127] Although Greenstein envisioned sharing the teaching load, he also proposed that a one-term course on radio astronomy be added to the standard curriculum in 1955. Greenstein ended up teaching most of this course alone, with occasional help from visiting researchers.

At Caltech, as at Manchester and at Harvard, pedagogy preceded both new instruments and new research. Only in 1954 did DuBridge decide that Caltech's radio astronomy program required a new dedicated staff member and a firm commitment to a type of radio astronomy instrument well suited to the demands of the world's foremost observatory. With some early funding from the Office of Naval Research, DuBridge offered William Pickering of Caltech's Jet Propulsion Laboratory the job of directing the new radio astronomy program. Had the director of the Jet Propulsion Lab not "suddenly . . . resigned," Pickering may have become a working physicist directing a radio astronomy program. Pickering seems to have been the leading candidate to direct the new radio astronomy program for the same reason that Ira Bowen had assumed the director's job of the Combined Observatories a decade earlier. The choice of Pickering re-emphasized Caltech's commitment to interdisciplinary work, and he probably would have taken Caltech radio astronomy more toward complex-style science than toward community-style science. But Pickering had already been courted by the Jet Propulsion Lab to assume its directorship, and he turned DuBridge down.[128]

DuBridge sought talent abroad. At some point in the early 1950s, he had discussed with Edward Bowen the possibility that the Australian radio astronomer John Bolton might join a Caltech radio astronomy program. Bolton, "quite well known to the astronomers and radioastronomers" all over the world, had come to Australia's Radiophysics Laboratory just after the war and had been an early member of Pawsey's solar noise team. Independently, however, he had sought and found the earliest source of cosmic noise in the Crab Nebula, a discovery that had caused his status within the emergent radio astronomy community to skyrocket. By the early 1950s, Bolton's professional relationship with deputy chief of Radiophysics Joe Pawsey, had deteriorated, while in chief Edward Bowen he had found a kindred spirit who shared his interest in large instruments. Upon visiting Mount Wilson and Palomar and working closely with Rudolph Minkowski and Walter Baade on identifications, Bolton had impressed the Combined Observatories' astronomers. And so in late 1954 DuBridge approached Bolton with a proposal that he initiate "a program of research in radio-astronomy [at Caltech] in collaboration with the members of our physics and astronomy and electrical engineering division on the campus and also with the astronomers of the Mount Wilson and Palomar Observatories."[129]

DuBridge enthusiastically recruited Bolton. DuBridge's initial proposal to the Office of Naval Research, for example, "expected that there would be very close relations between the work in radio astronomy and the work carried on at the Mt. Wilson and Palomar Observatories."[130] DuBridge declared that Bolton's expertise would help Caltech to gain funds for building a giant radio telescope from the ONR.[131] Two further reasons Bolton appealed to DuBridge and the Caltech astronomers were that he had intellectual expertise and that he supported the social arrangements of the new community of optical and radio astronomy. Bolton's willingness to seek out Minkowski and Baade, his ties to Edward Bowen, and his relationships with Australian and English radio astronomers all weighed in his favor. Bolton publicly celebrated the likelihood that Caltech's would be "the first joint optical-radio project in Radio Astronomy."[132] His new colleague Jesse Greenstein noted the necessarily "close connection with the Observatories" but also cautioned that "in a sense the project [would] be quite independent."[133] Administratively, Caltech's program would be under the auspices of the Division of Physics, Mathematics, and Astronomy, whose chairman, Robert Bacher, may have thought less of the new field than Caltech's president did. Though Bacher concurred that there would "be major contributions from radio astronomy," he added a caveat: "these are going to be supplementary to the results obtained by the big telescopes."[134] Collaboration, close relations, even joint projects, yes, but radio astronomy always would be "supplementary" to optical astronomy. Radio had made substantial inroads into the astronomy community, but, as in many marriages, there remained inequality between the partners. Still, the dowry grew nicely. Caltech received an assurance of support from the Office of Naval Research in 1955, but the ONR wouldn't commit to a large telescope project from the outset. Even Caltech, it seems, when it offered the ONR foreign expertise, international cooperation, and an interdisciplinary program of a fundamental nature, couldn't immediately secure the abundant funding the physicists had come to expect. DuBridge would have to settle for less grandiose instrumentation.

Bolton received initial funds to establish a preliminary instrument, which he proceeded to build within sight of Palomar. One cannot find a more telling image of the new meaning of astronomy. The dome of the 200-inch may rise above the small radio dish, but it is the new radio telescope that dominates the photograph. The dish's place within sight of Palomar evokes the close ties of the new radio astronomers building their telescope and the older traditional astronomers a short walk away. Superficially the convex of the dish is the inversion of the concave surface of the dome, but in reality Bush's analogy of the twin convex surfaces of dish and mirror holds true. Bolton may be only just completing his dish's construction, but his work is being done only a few years after the Palomar dome itself was completed—a dome that covered the greatest optical telescope then in existence. Together, of course, they are the new shape of the astronomy community.

FIGURE 3.2
The Caltech "32-foot reflector," June 1956. From Lee DuBridge Papers, 10.20–30. Reproduced courtesy of California Institute of Technology Archives.

One day near the end of 1952, Jesse Greenstein sat in his office in Pasadena considering radio and astronomy. A member of the foremost observatory in the world, Greenstein had cunningly kept one eye (or ear as it were) on the potential impact of the new technologies that had emerged during the war. For astronomers, the most immediately obviously applicable technology was that of Germany's ballistic rocketry program. Any interested astronomers, however, found it difficult to gain priority against military imperatives, and a veil of secrecy soon shrouded further efforts. Another path lay open to the astronomers. Radio techniques suffered few of the restrictions of the military, and they had the added advantage that the atmosphere didn't block radio wavelengths coming from space. Though Greenstein had been an early supporter of rocket-borne astronomy, he considered the technology unreliable. He soon concluded that ground-based platforms were preferable.[135] At least until the successful launch of Sputnik, in 1957, astronomers generally agreed with him. Despite all the claims being made about the potential of rocket-borne astronomy, Greenstein noted that radio had been responsible for identifying regions of neutral hydrogen with its telltale "21-centimeter" line, that other lines might soon be discovered, and that it would then become possible to detect optically unobservable celestial objects.[136] Even more exciting was the discovery that, although the Puppis A and Cassiopeia A sources were galactic, the Cygnus source, Cygnus A, was identified as extragalactic.

Thus, in less than a decade, an entirely new class of astronomical objects had been located, and, with young British and Australian radio physicists working alongside astronomers in California (while a radio physicist taught astronomers about electronics in Harvard), a new community had been forged. Soon there would be giant radio telescopes. With the new standards of "telescope" and of astronomical data, all astronomers—optical and radio astronomers alike—had become disciples. No longer did the visual hold a sinecure over astronomy. Greenstein's handwritten notes for a talk show that he debated with himself the merits of recruiting "physicists vs. radio engineers vs. astronomers."[137] After journeying to the Mount Wilson and Palomar observatories, young radio physicists became allied with the premier big-telescope astronomers of their generation and became the first radio astronomers. The radio physicists brought their radio data to the astronomers and asked for optical correlations. The results were astounding. When Walter Baade at Palomar Mountain looked closely at "the structure of this spiral object," it appeared to be "two extragalactic nebulae in actual collision."[138] With hydrogen's 21-centimeter signature and a galactic pileup making headlines in the 1950s, and with the discovery of the cosmic microwave background and quasars in the 1960s, radio astronomy rapidly came to occupy a powerful place within astronomy—a place it held until the Hubble Space Telescope reasserted optical astronomy by taking the telescope above the atmosphere.

In short, astronomy looked different at 21 centimeters. The telescopes that had charted the heavens ever more finely for 400 years had never seen beyond the visual. Radio could look at other wavelengths to reveal the hidden structure of the galaxy and could enable astronomy to see inside stars and listen to the noisy heavens. Before 1951, Australia's Joe Pawsey and Cambridge's Frances Graham-Smith went to the optical astronomers with humble requests for visual searches for objects that might correspond to their radio sources. Oh yes, the astronomers were all very welcoming, as one might humor an undergraduate wanting to work in a research lab; they were shown the wonders of the world-class telescopes, listened to kindly, helped with their little projects, and sent home. But within a year after the 21-centimeter observations, one of the oldest observatories in the United States, the Harvard College Observatory,

had new courses, new students, new instruments, new staff members, and new patronage, all in radio astronomy. Likewise, the self-proclaimed "world's center for astronomy" recruited a new radio astronomer, taught electronics to its graduate students,
and built a radio dish a stone's thrown from the 200-inch Palomar telescope. From
the East Coast to the West Coast, and from old to new, the culture of astronomy
changed as graduate curricula added technical courses on electronics, and as courses
from other departments became recognized as necessary to a future astronomer's
education.

When Robert Hanbury Brown entered the PhD program at Manchester, and when
John Baldwin did so at Cambridge, neither man knew any astronomy, but each reckoned that the subject would be easy enough to learn. For that matter, no one in the
entire Australian group knew any astronomy; neither did Harold Ewen or Edwin
Purcell, though they revealed an entirely new sky for astronomers to investigate. To
become radio astronomers, however, the radio physicists had to learn astronomy, carefully crafting their data in visual form and accepting a standard of error much finer
than before. At the same time, the three major radio astronomy groups—those at
Manchester, Sydney, and Cambridge—fought for students. Students became part of
the moral capital of science in a major way during the Cold War era, if only because
of the government largesse that was connected to graduate student recruitment and
training. Astronomy meant cooperation, between people, between institutions, between
disciplines, between countries, and even between masters and disciples. In 1956, when
the Dutch astronomer Jan Oort wrote to a recent Harvard graduate named Thomas
Menon about his proposed program of observations on the Orion region, he was
concerned "to avoid duplication." Menon, having just completed his PhD, was more
than happy to "co-operate" with Oort.[139] Thus, by the mid 1950s the social and
intellectual foundations of radio astronomy were essentially complete.

Once a new conception that included radio as a coherent part of astronomy was
securely in place, astronomers could build bigger and bolder instruments to penetrate
the heavens still further. The least of these new instruments was Harvard's new 60-foot
radio telescope, but its "very smooth mesh" became a point of pride for the Harvard
group. Observing such developments at his alma mater, Jesse Greenstein joked with
John Bolton that Caltech should also build a 60-footer (as Merle Tuve seriously recommended), asking "What's a few hundred thousand dollars among friends?" Greenstein's only real criticism of the Harvard instrument had to do with what he saw as
the over-engineering of the structural members, which he thought gave "the whole
device . . . several orders of magnitude of surplus strength."[140] He didn't doubt,

however, that the 60-foot instrument would help keep Harvard's observatory competitive with other radio telescopes around the world.

The instrument served as a metaphor for the new community itself. Astronomy had regained its strength by reasserting a commitment to broadening its intellectual and social community through cooperation rather than competition, whether disciplinarily, nationally, or between masters and their disciples. In 1954, one Australian explained this to a radio audience as follows: "In ordinary Astronomy, we gaze out, either by eye or with a powerful telescope, at the universe around us. . . . In this new science of Radio Astronomy, we sweep the skies with our radio telescopes—which are nothing more than very sensitive radio receivers connected to powerful radio aerials. [That] is what Radio Astronomy is: a new branch of Astronomy."[141]

Never was research so imperative in Australia—never has it been so costly. In order to conduct research, good accommodation, first-class personnel and equipment are needed, and not-withstanding the Cambridge tradition, the better the facilities the better will be the work.[1]

—Ian Wark, head of CSIRO's Division of Industrial Chemistry, 1951

Speaking at a 1955 meeting of the International Astronomical Union, Harlow Shapley, the former director of the Harvard Observatory, declared that astronomy had just experienced its most successful year. "Never in the long history of astronomy," he said, "has there been more building and planning of astronomical instruments than in 1954 and 1955, and rarely if ever has there been so much cooperation." Proudly reporting that several new optical telescopes had been built (notably one at the Lick Observatory in California), Shapley also mentioned the "large radio telescopes in England, United States, Holland, Germany, and Australia" and validated the integration of radio techniques into astronomical practice, the creation of the radio astronomers, and the burgeoning "internationalism" of astronomy.[2]

New disciplines, new disciples, and international cooperation had reshaped astronomy in a very short time. In only ten years, the international body of astronomers had included and transformed the antennas and laboratories of the radio physicists into the telescopes and observatories of the radio astronomers. Simultaneously, the conception of astronomy itself had changed along with astronomer's notions of their telescopes and of who might be an astronomer. By the mid 1950s, the radio astronomers stood poised to expand their telescopes still further. Shapley celebrated the vision of new giant radio telescopes.

Between the mid 1950s and the mid 1960s, five of the major groups of radio astronomers built large radio telescopes. Those telescopes would fix an interdisciplinary and international community in steel, cement, and electronics. In turn, the planning, funding, and operation of these new giant radio telescopes within radio observatories would solidify the moral economy of the radio astronomy community. This chapter looks at four of them: Jodrell Bank outside Manchester, Caltech's Owens Valley Radio Observatory, the Parkes radio telescope in Australia, and the Mullard Radio Observatory at Cambridge in England. Each of these was designed, funded, and built within ten years. Each telescope has its own story, of course. But when we look at the four concurrently (as the early radio astronomers themselves did) we see the historical development of a changing understanding of the telescope as an instrument, of astronomy as a scientific field, and of a scientific community as a social organization.

The American National Radio Astronomy Observatory will be taken up in the next chapter. In brief, it is the exception that proves the rule about the emergence of the radio astronomy community. As an avowedly "national" facility consciously designed to replicate the national facilities of the nuclear physicists, the National Radio Astronomy Observatory's first giant radio telescope failed to claim uniqueness and encountered substantial resistance to selecting a director solely on nationalistic criteria. In contrast with what had happened in physics, the limitations of "national" facilities for science were revealed. Governments might build scientific monuments, but without an international community of scientists and students they were empty.

In the four cases discussed in this chapter, community and instrument strengthened each other. Most obviously, Bernard Lovell and Edward Bowen built similar large and fully steerable parabolic dishes. Since parabolic dishes reflect incident radio waves to a single point of focus, Lovell's and Bowen's giant parabolas presented an obvious radio analogy to optical astronomers' understanding of a telescope. However, Martin Ryle issued a substantial challenge to that easy analogy between the radio and optical telescope. At Cambridge he built a large interferometer that added radio signals from two distinct perpendicular axes together electronically, which provided greater resolution but less sensitivity. Finally, at Caltech, John Bolton created one of the most successful radio observatories by innovatively combined these two approaches, placing a pair of large parabolas in an interferometer arrangement.

Considered together, the giant radio telescopes all evoked the international, inclusive, and cooperative radio astronomy community. The power of community was that it permitted, even encouraged, minor local differences. When each telescope has been considered only in isolation, the local differences have appeared significant. For example, Bernard Lovell's Jodrell Bank established the priority of the parabolic design and made radio astronomy recognizable as astronomy. Edward Bowen's Parkes radio telescope gazed upon the southern sky from Australia, and Bowen's young radio astronomers saw senior Californian astronomers as their fellows. Yet when the radio telescopes are considered together, it is evident that both Lovell and Bowen desired to bring radio astronomy and optical astronomy into a new all-wave astronomical community. Lovell and Bowen shared an international, interdisciplinary, and pedagogical vision of the community that culminated when the Australian radio astronomer Bolton powerfully combined parabolic dishes and interferometric techniques in Caltech's twin 90-foot dishes at Owens Valley. Many years ago, the sociologists David Edge and Michael Mulkay emphasized the divergent technologies of radio telescopes—especially the contrast between Martin Ryle (who eschewed the spectacle of the giant parabola and who built his mile-long interferometer outside a small market town on the East Anglia fens) and Lovell and Bowen (who favored giant parabolic dishes). In my account, technical differences between radio telescopes fade before the unity of radio physics' contribution to all-wave astronomy.

This chapter argues for the unity of the community even in the face of divergent national priorities and an inflamed conflict between parabolas and interferometers. What follows is thus a social history that ties together the creation of the measure and meaning of the new large radio telescopes and the stabilization of an all-wave astronomical community. The carefully constructed community was threatened by the differences between the two technical visions of radio astronomy (parabolas and interferometers) and between the commensurate social organizations. As Warren Weaver of the Rockefeller Foundation, one of the chief backers of the large Australian radio telescope, succinctly noted, radio telescope designs faced a major choice between "the compromise between the capacity to gather energy" (via parabolic dishes) and "the capacity to resolve" (via interferometric arrays). As Weaver explained, "the capacity, when receiving a signal from two sources which are very close together, to distinguish that one is in fact dealing with two signals from two separate sources, rather than a combined signal from a merged source. The larger the 'dish' is, the greater the energy-gathering power, and also the greater the resolution: but the cost also goes up very rapidly with the size."[3]

Weaver exposed one big question of the "big science" era: On what basis could increasing size be justified? Divisions between the supporters of interferometers and the supporters of parabolas threatened to pull the community apart. Yet the community agreed that large radio telescopes were needed to pursue research programs of identifying radio stars, mapping the radio heavens, and providing new windows to the heavens via radio wavelengths—especially the 21-centimeter wavelength of neutral interstellar hydrogen. In the end, the process of transforming local radio telescopes into global radio observatories overrode conflicts between supporters of parabolas and supporters of interferometers. The science of radio astronomy stabilized when Martin

Ryle firmly became a member of the community, John Bolton came up with a design that unified the two instruments, and the rebellious Australian interferometer crusader Bernard Mills was relegated to a physics department. Not only did the radio astronomy community survive the parabola/interferometer debate; it emerged stronger and more robust, its moral economy firmly built upon international and interdisciplinary cooperation.

A POSTWAR BRITISH SPECTACLE

The vision of giant radio telescopes originated in Manchester. Bernard Lovell's impressive performance before the Royal Astronomical Society in 1949 made radio physics data and instruments recognizable to an astronomical audience. Soon afterward, an influential Committee for Radio Astronomy sought to guide and financially support British efforts in radio astronomy, especially the provision of giant instruments and facilities. At its very first meeting, in February of 1950, the committee considered the question of how to support radio astronomy, and Lovell presented his audacious suggestion in full: a 250-foot-diameter parabola mounted so that it could point toward any region of the sky. It would be an icon, a white vision of postwar science. Moreover, many of the elements of Lovell's next-generation radio telescope would be replicated by most of the other giants built around the world in the next ten years. All but one followed Lovell's parabolic dish design, all gained funding and support from national governments as well as independent sources, and together they fixed radio astronomy as a constituent part of astronomy and of the global astronomical community.

The road to the world's first giant radio telescope began in April 1952 when Lovell received a grant amounting to about 335,000 pounds from the British Department of Scientific and Industrial Research and the Nuffield Foundation. According to the historian Jon Agar, the DSIR supported the project because of its "timeliness and promise." The unexpected swiftness and size of the grant in an era of British austerity stemmed from the "postwar economy of credit between British science and government" (credit based on science's contribution to the war effort), but also from the ease with which civil servants could envision the giant parabola as a symbol of national leadership.[4] Lovell certainly appealed to Britain's bruised nationalism of the 1950s when he offered to secure Britain's place in radio astronomy by building the world's largest dish.[5] A consensus quickly emerged in British scientific circles that Lovell's giant parabola held enormous promise for the future of science in Britain.[6] Sir Ben Lockspeiser, secretary of the DSIR, echoed Lovell's dream that a giant dish might "prepare a new map of the universe which would be even more complete than the one obtained by the great American astronomical optical telescope."[7] Nationalism required a national instrument to trump the instrument of some other nation, preferably a stronger one, yet ultimately consensus between instruments and standards was required; otherwise the comparison held no meaning.

National competition was thus merely an opening gambit. Once Lovell's proposal for a giant radio telescope hit the Royal Astronomical Society, the professional body of British astronomers, the measure and meaning of the instrument began to gradually change from a national declaration of prowess and leadership into a device marked by concessions and compromises. As Edward Appleton (a senior member of the British scientific community) remarked, the telescope promised to attack a list of problems in astronomy. Incorporating various designs and systems to address the expected list was challenging enough, but Appleton was "even more impressed by the possible uses of this instrument in fields of research which we cannot yet envision."[8] In other words, the telescope had to be designed for problems or even entire areas of science not yet imagined. Appleton's views were hardly isolated. After Sir Lawrence Bragg wrote to Martin Ryle in 1955 to ask him for a popular account of radio astronomy for an exhibition, Ryle overtly rejected any claims to practicality. Trying to be either helpful or sarcastic, Ryle wrote: "I think it is going to be a little difficult to get over the point of our exhibit to the general public, partly because of the few practical applications, and because the aims of astronomical research are always a bit obscure. I imagine there will be practical applications of radio astronomy; someone in America has already made a radio sextant."[9]

As a new science in the Cold War, radio astronomy offered no evident immediate benefits. Its value was in being a fundamental science. In the 1950s, potential fundamental discoveries held special appeal to private donors. The Nuffield Foundation became one of the single largest contributors to Jodrell Bank. Lovell had an ally in Sir John Stopford, who influentially and simultaneously enjoyed being both vice-chancellor of Manchester University and vice-chairman of the Nuffield Foundation. Even at the earliest stages of planning, Lovell knew that the cost of the parabola would exceed any potential DSIR grant, yet no amount of campaigning by Henry Tizard, Patrick Blackett, or Lovell himself could convince the DSIR to increase the allocation beyond the initial figure of 259,000 pounds. According to Lovell, any increase was politically unrealizable, since the DSIR was again operating under a Conservative Churchill government.[10] Within only a few years, extreme shortfalls in funding saw Lovell approach the Nuffield Foundation once more. The second time, in a deal that matched the DSIR's grant pound for pound, the foundation agreed to support the telescope even as the cost estimate ballooned. And balloon it did. Within two years, the cost of Jodrell Bank nearly doubled, from the initial request of 259,000 pounds to more than 450,000 pounds in mid 1953, and in mid 1955 it rose to 630,000 pounds.[11]

The rapid inflation in costs stemmed from the desperate attempt to provide one centerpiece instrument to fulfill multiple research needs. Yet those needs kept shifting even as the telescope took shape. At the outset, Lovell and his chief engineer, Charles Husband, decided that "it was [a] commonsense decision to take a wavelength of 1 [meter] as the shortest working wavelength of the telescope at which its beamwidth would be ±½°."[12] The choice of that wavelength stemmed from Robert Hanbury Brown's experience of studying cosmic noise at wavelengths on the order of 1 meter with the transit telescope and from the research problems that Lovell envisioned, particularly cosmic ray showers. Moreover, the 1-meter wavelength permitted economizing on the mesh surface of the telescope. The decision, however, plagued the telescope's development and became a focal point of every other radio astronomy group's critique of Jodrell Bank throughout the 1950s. Within a few months, Lovell and Husband had altered the design in the wake of American, Dutch, and Australian work on detecting neutral interstellar hydrogen at the shorter wavelength of 21 centimeters. In Australia, Edward Bowen thought the inability of Lovell's telescope to operate at the 21-centimeter hydrogen-line frequency was "a serious mistake."[13] Lovell too accepted that. He and Husband desperately attempted to meet the tolerance for 21-centimeter work, resurfacing the central 100 feet of the Jodrell Bank Mark I. That was only the first of many compromises. Even that concession created a substantial extra wind load on the central section, which in turn required extra steel to reinforce both the dish and the towers, precipitating yet another financial crisis as British steel prices rose.[14] The changes affected the whole design, resulting in "insufficient stiffness in the main axis, thermal expansion, and icing."[15] And with every design change, the cost soared.

Because the new Jodrell Bank telescope would operate at radio wavelengths, rain, cloud, and dew would not be problematic for it, in contrast with its optical counterparts. On the other hand, interference from the local radio environment necessitated another series of design changes and, eventually, political campaigns. Everything from the growth of local radio and television stations to emissions from vehicles would have to be controlled to ensure the integrity of the telescope and, by extension, the integrity of new British astronomy. The pre-eminent Dutch astronomer Jan Oort queried how Lovell planned to limit interference. Lovell promised that the design would reduce the "intensity of the interference by 400 times" by using "a focal plane design" reducing "the spill from the primary feed over the edge of the dish" and "by siting the new instrument" at a greater distance from "the nearest road."[16] Yet, as Agar rightly points out, the spectacular nature of Jodrell Bank severely limited the instrument's isolation from the environment. Indeed, it was supposed to be seen by passing motorists, who, it was hoped, would offer support or donations.

With Lovell overcommitted to design changes and to fundraising, much of the continuing research work at Jodrell Bank fell to Hanbury Brown and the transit telescope, which remained in service well into the 1950s. Using the transit telescope in combination with a smaller mobile dish, Hanbury Brown created a flexible interferometer. Still, even at a separation of 20 kilometers, three of the sources remained unresolved. The source's diameter was less than 12 seconds of arc, whereas the arc diameter for the Andromeda nebula was measured in minutes.[17] The nature, the distance, and the size of the objects remained outstanding problems for the Manchester radio astronomer and his two graduate student assistants for some time. Unable to identify the sources locally, Hanbury Brown sent a list of small-diameter sources to several major optical observatories and awaited their replies. The failure to identify most radio sources with any class of visible objects "led to a concentrated effort to find out more about the actual diameters of the radio sources," Lovell noted. The interference methods used by both the Sydney and Cambridge groups placed a limit on the separation of the elements and therefore a limit on the resolution achievable. Hanbury Brown's innovative technique supposedly solved the problem of how far apart the elements of an interferometer could be placed. Lovell's announcement to the astronomical community that "Hanbury Brown has recently designed a new type of interferometer in which the signals from the two aerials are compared after rectification and in which no limit is placed on the possible separation of the aerials" came just at the time when the costs of Lovell's giant parabola doubled.[18] In other words, just as the giant parabola began to take shape, an elegant and simple solution called its size into question. The Cambridge tradition had scored a point.

A GIANT RADIO TELESCOPE FOR THE SOUTHERN HEMISPHERE

Lovell's proposal for a giant radio telescope awed Edward "Taffy" Bowen of the Radiophysics Laboratory in Sydney. Lovell's vision having gained support, Bowen began his own campaign to erect, in Australia, a radio telescope at least as big as Lovell's. The prospects of a giant radio telescope in the southern hemisphere depended on the privileged position of Australia for galactic studies, as well as on the unique ability of radio astronomy to reveal the universe's most abundant element, hydrogen.

Radio astronomy at 21 centimeters promised "nothing less than a portrait of the Milky Way galaxy," Bowen explained to Caltech's president, Lee DuBridge, and the race to resolve the shape of the galaxy demanded that "some portion at least of a giant telescope must work on 1420 Mc/s."[19] For nearly four years, Bowen fixated on the idea of a giant radio telescope for the southern hemisphere.

Bowen gained the necessary funding in large part because he was the linchpin of the new international radio astronomy community during the 1950s. He remained on intimate terms with Merle Tuve, Vannevar Bush, and Lee DuBridge, non-astronomers but members of the cadre of American postwar scientific leadership that exerted great influence on American science policy. He regularly visited and corresponded with the British radio astronomers Bernard Lovell, Robert Hanbury Brown, and Martin Ryle. As it turned out, the Australian government supported Bowen's grand vision, but only long after one of the Rockefeller Foundation's last examples of private philanthropy toward astronomy. Warren Weaver of the Rockefeller Foundation made an exception for a Cold War policy against large instruments for the physical sciences and awarded $250,000 to Bowen's radio telescope project in 1955. (Of course, Bowen knew Weaver as well.)

Bowen's vision of a giant radio telescope for the southern hemisphere took shape between 1952 and 1956. Unlike Lovell in Manchester, Bowen integrated the telescope into a broader "Radio Observatory" in Australia from the outset.[20] An Observatory would address two matters that were especially important to radio astronomers in the 1950s: determining the sources of cosmic noise and investigating how those sources were distributed. With those priorities, Bowen's Radiophysics Laboratory abandoned research on the ionosphere, on meteors, and on auroras; it also abandoned radar echo research on the moon and other planets. Bowen told the nuclear physicist Mark Oliphant that he viewed Lovell's anticipated work on long-wavelength radar echoes from the sun and the planets as an "interesting technological achievement" that "might be quite disappointing from a scientific point of view."[21] Solar work continued both at meter and centimeter wavelengths, but it was considered dismissively "routine." New work was more in the vein of the Australian researcher Frank Kerr's work on hydrogen-line radiation. Interstellar hydrogen work promised progress on dust clouds, on dark nebulae, and on the structures of galaxies.[22]

The campaign for a radio observatory in the southern hemisphere opened at the 1952 meeting of the International Radio Science Union (URSI) in Sydney. URSI's president, Sir Edward Appleton, argued that Australian science would no longer be isolated and would no longer function solely for local purposes once a giant radio telescope had been built. Australia would enter the international scientific arena and would properly contribute to science.[23] The Australian government, however, was not to be as easily swayed as the British. Nor, despite efforts, did a local private champion come forward to support an Australian radio observatory. Lovell and Jodrell Bank both helped and hindered. The reality of Jodrell Bank gave Bowen a figure to aim for, a size to propose, and, perhaps most important, a reality that he could point to.[24] Then again, Bowen's southern hemisphere radio telescope would have to be comparable to Jodrell Bank and Palomar Mountain yet distinct from them. Bowen constantly reiterated how the Australian radio telescope proposal remained "substantially different" from the British model.

The Australians concentrated on utilizing a new design and less material to produce greater accuracy at lower cost. John Bolton, Bowen's protégé, faithfully supported the concept of a large dish—a fixed dish, some sort of rolling barrel, or even a Manchester-style "paraboloid of revolution." Bowen's deputy Joseph Pawsey noted that it was Bolton's aim to construct an "aerial so big" that it would "sort out the complex distribution [of overlapping sources] by virtue of its resolution."[25] What was needed was a rethinking of the support structures that would address the stress and tolerance problems. Several variants were dreamed up; two of the earliest were a dish "Supported at Several Points [by] (Hydraulic Jacks)" and a "reflector in the form of a long cylinder (1,000 ft) of width 200 ft . . . supported on the ground and steered ±60° by rolling." The barrel design addressed several engineering problems that continued to plague Bowen's dream of a single fully steerable parabola, particularly the weight of such a structure and the problems associated with "maintaining the parabolic surface to the required tolerance."[26]

To Bowen, it was clear that Lovell's strategy was the correct one. Radio astronomy had to be established in parallel with optical astronomy; a 200-foot radio telescope would complement the 200-inch optical telescope. Yet the public doubts of the Australian government reflected private misgivings within Bowen's staff. One of the younger Australian radio astronomers, Bernard Mills, took the opportunity to advocate interferometers—radio telescopes formed by "crossed arrays of aerials" rather than giant dishes.[27] Ever mindful of the Cambridge tradition, Pawsey believed that the multiplication of smaller instrumental set-ups and smaller working groups was still likely to yield new discoveries. Bowen's superior, Frederick White, acknowledged the danger of spending a vast sum of money on a single instrument: "no doubt . . . once one does build a large aerial system one is committed to a defined programme for a long period."[28] Bowen debated the appropriateness of a large telescope with Pawsey for months, and became increasingly irascible. In March of 1953, he expressed to White the opinion that Pawsey "knows all of the reasons why we should not have

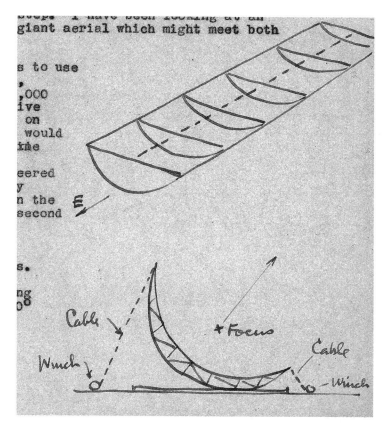

FIGURE 4.1

A "cylindrical paraboloid" design concept for a large antenna illustrated in Edward Bowen's letter to Frederick White dated October 22, 1952 (Australian Archives, series C3830, A1/3/11/1). Reproduced courtesy of Australian Archives.

one. This, of course, is exactly the way to put our feet in the grave as far as radio astronomy is concerned."[29]

With the plans for the giant radio telescope taking shape, Pawsey lamented the end of the laboratory era for radio astronomy. Radio astronomy, he said, had begun "moving towards the observatory procedure where complex equipment is used by a succession of observers to investigate explicit problems." The tradition of the Cavendish Laboratory centered on small groups building limited apparatus and seeking to maximize the research potential of any experimental set-up.[30] In contrast, Bowen's conception for a giant radio telescope in the southern hemisphere sought the complete adoption by Australian radio astronomers of "observatory" practice, as opposed to "laboratory" practice. The Cambridge tradition was messy, haphazard, and cheap. Bowen's vision of a large radio observatory was modern, planned, and expensive. A Cambridge laboratory might be a small group of researchers in a back room; the staff of a giant southern-hemisphere observatory would participate in a global radio astronomy community. Bowen, with his American connections, was enamored of the modernism of the American universities that were rushing to build large instruments. Certainly Bowen and White were in the vanguard of Cold War–era Australian science, which was distancing itself from the Cavendish tradition and turning toward more American styles of organization. Ian Wark, head of CSIRO's Division of Industrial Chemistry, summed up the move away from Australia's traditional adherence to the Cavendish tradition: "Never was research so imperative in Australia—never has it been so costly. In order to conduct research, good accommodation, first-class personnel and equipment are needed, and not-withstanding the Cambridge tradition, the better the facilities the better will be the work."[31] That was reminiscent of the speech that Australia's Prime Minister, John Curtain, had delivered after the fall of Singapore in 1941—a speech in which he had appealed to the United States rather than to Britain for military assistance against the Japanese.

The differing conceptions of "laboratory science" and "observatory science" help flesh out the picture I have been developing of two competing styles of science in the Cold War era—"complex-style science" and "community-style science." Consider the following episode: During the Cold War, as much in Australia as in the United Kingdom or the United States, any connection to national defense was the obvious route to gain funding for science. Bowen's superior, Frederick White, knew full well that selling the radio telescope project to the Australian government would entail promising practical outcomes at some level. White thus pragmatically asked Bowen "Is there any chance of linking this project with Defense?" If the radio telescope were "coupled with the very practical and important problem of long range aerial warning for the Air Force," it "might well have a hope of success."[32] Bowen cautioned that radio astronomy and radar had "tended to diverge." As Bowen outlined to White, "the converse argument is in fact the stronger one. If we were building a long cylindrical aerial for defence research there is no doubt it could be quite useful for some aspects of radio astronomy. If we were building one of circular section for radio astronomy, we could not honestly claim that it would be very useful for defence research."[33] Here is one of the most revealing moments in the creation of a giant radio telescope, indeed in the creation of the character of the radio astronomy "community." "Community-style" science remained distanced from military priorities because groups like the Australian radio astronomers simply admitted that they couldn't honestly justify any defense application for radio astronomy. Moreover, the Australians weren't alone in resistance the easy lure of military priorities to gain funding. As Jon Agar notes, in early 1954 "Robert Cockburn, then Scientific Advisor to the Air Ministry (and ex-TRE), considered there was "no direct defence interest in the use of the Jodrell Bank telescope."[34] In short, their inability (or unwillingness) to invent practical military applications for radio astronomy united the Australian and British radio astronomers further into a community.

The radio astronomers' unwillingness to transform their science to suit military needs governed their pursuit of community-style rather than complex-style science. No doubt an argument could have been made for the funding of defense-related radio astronomy. Since the work of Paul Forman, considerable historical attention has been spent on evaluating the extent to which science was shaped by the interests of the military-industrial complex. The case of the radio astronomers takes us in a novel direction, showing that defense-related priorities were not necessary to the building of larger scientific instruments. Scientists could choose not to reshape their projects to appeal to the military. For example, in Australia, Edward Bowen desperately sought funding for his giant radio telescope, but not so desperately that he would adjust his idea of a telescope. Consequently, Bowen, like Lovell, passionately advocated open, disciplinarily inclusive, internationally cooperative science—in other words, community-style science.

Without any clear military purpose, a familiar search for patronage ensued. In rapid succession, Bowen sent inquiries to Henry Tizard in Britain, to Vannevar Bush in the United States, to his own CSIRO masters, and, as always, to Lee DuBridge. As it turned out, Bush told Bowen about the existence and applicability of the large radio telescope project for one of the Dominion Grants from the Carnegie Foundation. The Dominion Grants had been sitting idle for some time during and after World War II, but were earmarked for large innovative scientific projects within the British Commonwealth. Bowen jumped at the opportunity, and his proposal reached the Carnegie Foundation in early 1954.

Vannevar Bush thought that the proposed location of the giant radio telescope in the southern hemisphere gave Bowen an excellent chance of gaining Carnegie support. Still, Bush cautioned, the Carnegie Foundation had moved away from funding the physical sciences, which now received a "heavy emphasis of government subsidy."[35] In fact, the Rockefeller Foundation and the Carnegie Foundation, once stalwart supporters of astronomy, had already shifted much of their support to more "attractive" areas (e.g., biology), and this seemed to leave new areas of the physical sciences, including radio astronomy, out in the cold. Bush's warning underscores a division between many government-supported Cold War "complex" sciences, such as nuclear physics, and more "community" sciences. To receive beneficent support from the philanthropic foundations, "community" sciences couldn't simultaneously depend on government money; instead they would have to advocate an international scientific community far removed from nationalist agendas. This was, in fact, what happened. When Bowen first approached Vannevar Bush about Carnegie sponsorship for the southern hemisphere telescope, for example, he made it clear that any Australian instrument project would not compete with any plans DuBridge had at Caltech: "On the contrary, I would like to think of it very much as a collaborative effort in which the work in the southern hemisphere was complementary to that in the northern. The close association we already have would ensure this from the outset. I would hate to think that any approach to the Carnegie Foundation on our part would adversely affect the chances of a similar device, or a better one, being built in California."[36]

Bowen also capitalized on the interest that Jodrell Bank had generated, and made deft use of a "strong body of opinion" within the American radio astronomy community that another giant radio telescope in the northern hemisphere would be "wasteful." "I am watching the situation closely and giving it a shove when needed," he wrote to Sir Richard Casey, Australia's Minister for External Affairs and Minister in Charge of CSIRO.[37] Bowen would be one of two Australians presenting papers at Jesse Greenstein's conference on radio astronomy in 1954. Bowen sought a month's leave from CSIRO to attend the conference, and entertained additional invitations from Caltech and from MIT. Frederick White granted him two months. Lee DuBridge believed that Bowen's presence in Washington was "very important."[38] Bowen believed that he and Mills received an "exceptionally good hearing" from the international crowd and claimed that together they "provided more meat than any other country."[39]

Bowen garnered the backing of astronomy's international body, the International Astronomical Union. In an open letter, the IAU's president, Otto Struve, and its General Secretary, Pieter Oosterhoff, praised radio astronomy for "yielding some of the most interesting developments in astronomy." After citing the "beautiful" observations of interstellar hydrogen as an example, Struve and Oosterhoff insisted that "the further development of the subject required much more detailed and precise observations which can only be obtained with much larger and more costly instruments." The radio and optical community agreed about the necessity of larger instruments, but Struve also weighed in on the choice of a radio astronomy instrument. Clearly, optical astronomers viewed the parabola as a recognizable analogy to optical telescopes. In turn, the international optical astronomy community gave tacit endorsement to parabolas. "The giant parabolic radio telescope," Struve and Oosterhoff claimed, "stands alone because it is the only known method obtaining the combination of high sensitivity and high resolution required in the outstandingly important shorter-wavelength region of the radio spectrum. This instrument is also exceedingly versatile so that it can be applied to different new problems as the need arises."[40] And Struve emphasized that the southern hemisphere sky, containing some of the most interesting regions of the Milky Way, was beyond the scope of Lovell's Manchester dish.

It was a testament to Edward Bowen's extensive networks and the strength of the radio astronomy group in Sydney that in May of 1954 the Carnegie Foundation gave the Australian CSIRO Division of Radiophysics a grant of US$250,000, representing approximately one-fourth of the estimated cost of building a giant radio telescope. Frederick White claimed that Bowen's visit to the president of the Carnegie Foundation in April was directly responsible for that institution's support. Merle Tuve viewed the Carnegie grant as a sign of the "vigorous support and approval" among Bowen's "scientific friends" in the United States. Vannevar Bush said he was "glad to put in a good word" on behalf of Bowen and the Australians, but it was evidently unnecessary; Bowen had already sold the Carnegie Foundation on the idea.[41] International community, scientific friends, a recognizable astronomical instrument of a new type, and a science not related to national defense priorities now became the model for the growth of facilities for radio astronomy.

In Australia, the celebrations were short-lived. Almost immediately, White and Bowen had to begin to secure matching funds from within Australia, as the Carnegie Foundation stipulated. One might expect that a substantial grant from a major philanthropic foundation would have been testament enough to the significance of the Australian contribution to a new field of science, but evidently not. The Australian government appeared especially unsupportive; Frederick White lamented to Henry Tizard in London that "there is little that one can present of a practical nature which might appeal to the imaginations of the members of Cabinet." Paralleling Lovell's experience in Britain, White thought the only way the Australian government might be swayed was "on the national prestige value" of a radio telescope.[42] The physicist Mark Oliphant, certainly the best-known Australian scientist at the time, agreed that the Radiophysics Laboratory's proposal would gain the appropriate support only "as a national undertaking and as a matter of national prestige."[43] Oliphant drew on his own experience of extracting money out of the Australian government for nuclear science. His claims again suggest that the nuclear sciences beat the drum of nationalism solely to gain bigger instruments.

Seeking powerful allies, Bowen asked Vannevar Bush to come to Australia. Bush expressed regret that he wouldn't be able to journey to Australia to personally implore business leaders, policy makers, and government officials to increase their support of the giant radio telescope: "I realize, I believe, what the general problem is in Australia, and I have met similar problems in industry, of course, on a very much more minor scale. Some of us know that it is impossible to carry on applied research in a thoroughly effective manner without the presence of basic and fundamental research with it or alongside it." The search for only applied or practical science was ultimately flawed, Bush told White. True scientists would eventually withdraw from the hollow pursuit of applied research. Bush's utopian vision centered on "community" science:

The urge to add in a creative way to the general understanding is very strong among individuals, and if there is not an outlet for this urge those individuals who feel impelled most strongly thus to contribute to the grasp of their fellow men will go elsewhere. If it is to be really effective even in applying science for practical ends, needs to have in its midst those who are reaching far ahead in their thinking and who are building the foundation for the applied work of a later generation, or even building a foundation so that man may better grasp his position in the cosmos and reason more effectively about matter of the spirit.[44]

Such high-mindedness may have been in the minds of some scientists, though one doubts that they dwelt much in either Bowen's mind or White's. Oliphant dismissively derided such whimsical ideas. In the context of Australian science in the 1950s, the reality of seeking any support from the Australian government rested on anticipated practical outcome for the nation. Toward the end of the war, White, like many scientists, had embraced the notion that fundamental science must be pursued either for its own sake or to underwrite future applied science. To that end, he had been unstintingly supportive of Bowen and Pawsey's radio astronomy research, even in the face of opposition from the local scientific establishment. "I have done my best to encourage Bowen and his colleagues to get into one or two quite fundamental lines in physics," he patiently explained to Vannevar Bush, "in spite of the fact that CSIRO generally likes to foresee some fairly reasonable application of its work. It is difficult to point to any even reasonably long term application of radio astronomy, but we have always argued that such fundamental work as this must, in the long run, be of practical significance to a country such as Australia."[45]

Sure enough, when the proposal for a giant radio telescope went before the Australian cabinet in 1955, the Minister-in-Charge of CSIRO, Lord Casey, pointedly asked the Secretary of the CSIRO Executive about "possible practical outcomes." Ideas abounded, of course, though White jovially reported to Joe Pawsey that Minister Casey had "given up the idea of tracing flying saucers." Even the CSIRO Secretary remained uncomfortably conscious of the fact that radio astronomy really didn't have any foreseeable practical outcomes. He pressed Pawsey continually for any "new angles" beyond "talk about fundamental research on the ionosphere leading to radar." As a last resort, the Secretary conceded, "we can, of course, use the Lord Kelvin touch (what is the use of a new born babe!)."[46]

While Bowen became "despondent," the Australian government remained "apathetic," and "people in industry" steadfastly believed that the project was the government's responsibility; CSIRO managed to raise merely 5,500 pounds from Australia's top twelve companies.[47] While Casey hoped that a "broader program of appeal" might raise some more funds, he faced the reality that no amount of campaigning would garner any substantial sum from the Australian public.[48] Australian industry and business would eventually donate less than 1 percent of the cost of the instrument. Casey always hoped to interest private industry in the project, especially after the Carnegie donation. He sent off a series of letters to "a few substantial individuals" to ask for support. His letter to G. J. Coles, chairman of a big supermarket concern, elevated matching Jodrell Bank in the northern hemisphere to the primary reason why industry should support the project. Secondarily, "it is a compliment to Australia that [the Carnegie Foundation] have made this proposal to Australia and not to other Southern Hemisphere countries." In a world suddenly divided into hemispheres rather than nations, simple nationalism sounded especially hollow. As the secretary of the CSIRO foresaw, proponents of the giant radio telescope resorted to Lord Kelvin's argument: "[T]he immediate purpose of the Telescope [was] to pursue fundamental research [and], as you know, most fundamental research turns out to have practical applications in due course." Casey proceeded to list several possible practical outcomes of the project, including increasing sensitivity in radio receivers, increased accuracy in direction finding, and investigations of the radio transmittance and interference characteristics of the ionosphere, and perhaps even some "military applications."[49] But the promise of practical benefits wasn't successful in luring either Australian industry or the Australian government.

With at least the Carnegie grant in hand, Bowen moved toward developing a design. Using a form-follows-function rationale, he stipulated that the "really important advances" in radio astronomy would come from hydrogen-line studies.[50] Hence, any radio telescope design required the greatest, and most exact, reflecting surface for 21-centimeter work. Bowen proposed reducing the overall size of the dish from 250 feet to 210 feet, but with the central 150 feet all smooth enough for hydrogen-line studies. Merle Tuve of the Carnegie Institution argued that the Radiophysics Lab should consider a smaller dish, as telescope size was related to engineering potential but not to knowledge production. Instead of 250 feet, "if you have a 170 foot dish with a good drive you will be preeminent in the field for at least ten years," he said.

There was little point in aiming for a "resolving power of 1/10° . . . worth millions" when a resolving power "of say 1/3°, which can be obtained with a good 110 foot dish" could be had for about US$600,000, slightly more than double the Carnegie grant. Tuve warned that the problems of tracking and surface deflection "would become "almost intolerable for a 250 or 300 foot dish, as per Jodrell Bank and Berkner." As we shall see in the next chapter, Tuve's particular criticisms stemmed more directly from his own ongoing fight with Lloyd Berkner, who wanted the "world's biggest" dish for the American National Radio Astronomy Observatory. Tuve couldn't see the sense in "an irrational ambition for the 'biggest.'" "Very little of the foundation rests on anything related to astronomy," he chided Bowen. "It is a kind of engineer's paradise."[51]

In late 1955, the Australian government decided to match, pound for pound, both the Carnegie grant and any addition money the Radiophysics Lab could acquire. Even with the size of the dish reduced, a substantial budget shortfall remained. To make matters worse, the United States' potential for supporting radio astronomy plummeted that year when Bowen learned "that the Associated Universities [were] going to build the biggest radio telescope in the world." "We just can't compete," Bowen conceded to the CSIRO's new chief, Ian Clunies-Ross.[52]

More than a year after the Carnegie grant, salvation came once again from the United States. After extensive lobbying from Lord Casey in his dual role as Minister for External Affairs and Minister in Charge of CSIRO, as well as continued advertising from Bowen and his American network, the Rockefeller Foundation donated US$200,000 to the giant radio telescope project. Warren Weaver, now well known to have steered the Rockefeller Foundation away from the physical sciences and toward the biological sciences, took unusual interest in Bowen's proposal for a giant radio telescope. From the records we have of the Rockefeller deliberations, it is clear that the foundation viewed radio astronomy as a very different enterprise from the majority of the physical sciences, which it no longer supported. The foundation received "three requests for major assistance in connection with the building of radiotelescopes" in August of 1955. "One of these is from the Bowen group in Australia, one from California Institute of Technology, and the third from Ohio State." John Kraus at Ohio State University had developed a novel design for a radio telescope with a reflecting screen. Kraus had been one of the early beneficiaries of funding from the National Science Foundation's Committee on Radio Astronomy, and was now looking for more. Expansion also dominated Caltech's request. Under Lee DuBridge, Caltech was expanding into radio astronomy to complement its optical program. And Edward Bowen (the only person Weaver mentioned by name) desperately needed the finding to realize his giant radio telescope. From Weaver's diary, we learn that Australia's geographical position was important, balancing the northern and southern hemispheres. Likewise, the fundamental nature of radio astronomy scored an important point in Bowen's favor for the Rockefeller Foundation, which condemned both industry and government for insufficient support. National affiliation played no part in the final decision. Weaver saw a grand endeavor:

I think there can be no doubt that radioastronomy presents, at the present moment in scientific history, a peculiarly interesting and, one might say, poetic possibility. This technique will certainly increase in a very significant way our knowledge of the universe in which we exist. It is clear that this new technique supplements in an exceedingly interesting and important way the evidence which can be obtained through various types of optical telescopes. From the geometrical nature of the case it is important to have first-rate installations in the Southern Hemisphere as well as in the Northern Hemisphere.[53]

Two weeks later, the Australian Radio Telescope was fully funded, half by private philanthropy from the United States and half by the Australian government. Design studies by Barnes Wallis (inventor of World War II's "Dam Busters" mission) and by the consultant engineering firm of Freeman Fox and Partners (who had designed the Sydney Harbour Bridge) were not completed until late 1957. Once the decision had been made to proceed with an altazimuth mount and an agreement had been reached that 210 feet was the maximum size the money would permit (the total cost being estimated at 500,000 Australian pounds), the building of the radio telescope proceeded over the next several years.[54]

As Bowen's giant dish took shape in a field outside the country town of Parkes, two of the younger members of his staff, Bernard Mills and Chris Christiansen, criticized the entire project and fought to gain substantial funding for their own giant Mills-Cross interferometer. The criticisms made by Mills and Christianson had to do with the dominance of parabolic radio telescopes, but also with the relationship between radio physics and optical astronomy that the parabolas represented. To understand their criticisms fully, however, we first have to consider the emerging champion of interferometer-style radio telescopes, Martin Ryle.

MARTIN RYLE AND THE CAMBRIDGE INTERFEROMETER

Lovell's use of the term "radio astronomy" to describe both radio and radar observations deeply dissatisfied Martin Ryle at Cambridge. Radio involved the passive reception of radio noise emissions, Ryle said, whereas radar required receiving active reflections from astronomical bodies, especially meteors. Ryle suggested that "radio astronomy" should be distinguished from "radar astronomy," even though both names could signal radio practices rather than "conventional astronomy at visual wavelengths."[55] As Lovell's giant dish took form, Ryle claimed that regular chartings of the sun and galactic sources were only the initial stage of the move toward an observatory. In contrast with the radar astronomy at Manchester, Ryle planned on pursing radio astronomy at Cambridge, which meant that Ryle "planned to continue and develop the 'observatory' aspect of the work." Like Bowen, Ryle understood that "radio astronomy" implied the construction of a "Radio Astronomical Observatory."[56]

Though Ryle and Lovell differed on the question of the form of the radio telescope (parabolic dishes versus interferometers), they fundamentally agreed that an observatory was the place for radio astronomy. Edge and Mulkay and Sullivan have given excellent explications of Ryle's method of aperture synthesis and its role in the 2C and 3C catalogue controversies between the Australian and Cambridge groups over the density of radio sources in the galaxy.[57] With those surveys the Cambridge group continued to hope "that the positions of a considerable number of sources may be found with sufficient accuracy to warrant a more extensive search for related objects."[58] The source controversy was a direct outgrowth of astronomers' initial interest in radio astronomy, the correlations between optical and radio objects, particularly of unusual types. The story thus far has refuted one of Edge and Mulkay's major claims: that the controversy grew out of Manchester's engagement in the astronomical world and Cambridge's relative isolation. We have seen that their struggles over the Cambridge 2C and 3C source catalogues emerged from Bernard Mills' and Francis Graham Smith's visits to Walter Baade and Rudolph Minkowski at Mount Wilson in the early 1950s and their extensive ongoing correspondence. In fact, the controversy centered on which group best integrated radio physics and optical astronomy to create radio astronomy—in other words, who was the most communal in a community-based science.

It is true that Ryle's use of aperture synthesis offered a critique of Bernard Lovell and his Jodrell Bank dish. The giant parabola offended the Cavendish tradition through waste, extravagance, and inefficiency. The parabola seemingly combined the weaknesses of previous methods rather than their strengths. As Ryle explained, "the further study of radio stars involves two basic requirements: the instrument must have a high angular resolution so that adjacent sources may be distinguished individually, and it must possess a large 'collecting area' in order to detect faint sources without an inconveniently long integration time." Ryle claimed that "conventional methods" to maximize both resolution and sensitivity were, in fact, "conflicting." The aperture synthesis method increased both the sensitivity and resolving power of the "radio star instrument."[59] For Ryle, Lovell had failed to create a large aperture in radio astronomy equivalent to an aperture for optical astronomy. The equivalent, Ryle explained, would give Jodrell Bank a resolving power of one-fortieth of a radian, or 1.5°; an optical telescope with a resolution of 1.5° would see the sun as "a diffuse blur . . . about three times its present size."[60] The critique itself was a stunning piece of rhetorical strategy. It highlighted the astronomical deficiencies of the Jodrell Bank telescope and created an argument for Cambridge-style high-resolution interferometers as the better analogy to the optical telescope. In other words, Ryle's major objection to Lovell came on astronomical grounds. As a radio astronomer, Lovell should be attempting to match the standards of the field—needless to say, well in excess of 1.5°. In contrast, Ryle's methodology partially stemmed from optical astronomy itself. Ryle attributed the principle of aperture synthesis to Albert Michelson's stellar interferometer, in which two mirrors spaced several meters apart acted like the edges of a large lens. Michelson's triumph, and Ryle's, was to "use measurements from the resulting interference fringes to achieve a much increased resolving power."[61]

For all that, aperture synthesis possessed none of the easy analogies to optical astronomy, nor the gleaming, towering white presence on the fields of Jodrell Bank. Interferometers spread out horizontally along the ground, and gave the appearance of large, intricate fences. In the American magazine *Sky and Telescope*, Lovell outlined the main differences between Cambridge and Manchester solutions to the problem of resolving power. Ryle at Cambridge used radio interferometers, he said, but "at Jodrell Bank, resolving power has been improved by increasing the physical dimensions of the aerial system."[62] Without substantially justifying the claim, Lovell's statement implicitly accepted that resolution was paramount in radio astronomy. Around this time, Martin Ryle had noted to himself that Lovell's "proposed instrument would not, in itself, lead to any increase in accuracy of positioning of the sources already known, over that obtained by interferometric technique."[63] Evoking the theme of spectacle, Ryle drew distinctions between size and elegance, between the single parabola and the interferometer. The advantage of the large antenna lay "in the ability to carry out accurate surveys over a wide range of wavelengths." As Ryle noted, a large parabola didn't fulfill the current demands of optical astronomers for more accurate positions of radio sources any better than his interferometer. "As far as can be seen at present, [Jodrell Bank] possesses no features which make it in any way alternative to the interferometric techniques for the delineation of the point sources."[64]

By mid 1954, Ryle considered his Cambridge group second only to Bowen's Australians. In spite of the large grants given to Lovell to erect the world's "largest parabolic reflector," Ryle believed, the significance of their survey and their innovative technological solution to radio star positions gave Cambridge the edge. Yet Ryle foresaw the eventual eclipse of Cambridge radio astronomy because, though Lovell's giant dish failed at many levels, one had to agree that its size alone made it significant. Looking to the future, Ryle saw Cambridge's "comparatively small grants" as ultimately limiting, even while though supporting scientific results and instrument design of "importance and originality." In addition, Ryle's interferometers had reached a stage, he sagely noted, where, despite great resolving power to initially locate radio sources, "a disadvantage" had emerged when distinguishing signals from adjacent sources. These two reasons combined to force Ryle to follow Lovell and Bowen's lead.

Flying in the face of the Cavendish tradition of small and elegant, Ryle proposed and eventually built a large interferometer-style radio telescope. The Mullard Radio Observatory commenced operation in 1958, before Australia's Parkes radio telescope and only shortly after Jodrell Bank. Ryle's was a "Michelson Pair" interferometer. As

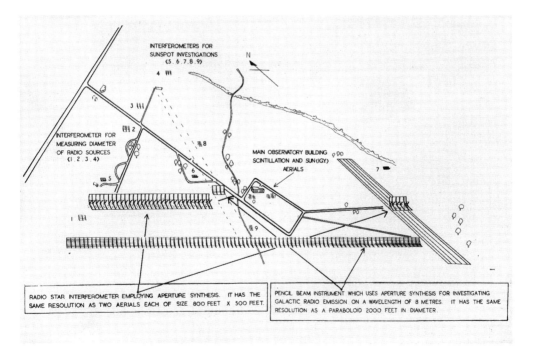

FIGURE 4.2
"Plan of Mullard Observatory Showing Positions of Radio Telescopes" (Box 7/14, Martin Ryle's Papers, Churchill College, Cambridge). Reproduced with permission of the Master and Fellows of Churchill College, Cambridge.

FIGURE 4.3
The Mullard Radio Astronomy Observatory at Cambridge. Reproduced with permission of
the Astrophysics Department of the Cavendish Laboratory of Cambridge University.

Ryle explained, by utilizing variable distances from observation to observation it
would allow determination of the "distribution of intensity across the source as a
Fourier synthesis." In combination with another interferometer running perpendicular
to the first, forming a T, the interference of overlapping signals "from neighbouring
sources" would "be largely eliminated" while still maintaining a "resolving power
appropriate to both dimensions of the T." In effect the design would again contribute
to the overall aim of the Cambridge group: "to improve the accuracy of the measured
position of several of the known sources."[65] Ryle's instrumental direction was unique.
When Ryle looked at the programs of Manchester, Harvard, and Australia when
evaluating his own plans for a radio observatory, he decided against "the construction
of a single large general-purpose type of instrument." Cambridge's Radio Observatory
would, rather, construct "a number of specialized instruments, each designed to investigate a particular type of problem. In this way it seems possible to build more powerful instruments at considerable smaller cost."[66]

Beyond the form of the radio telescope, however, the men at the Cambridge radio observatory maintained the same commitment to the ideals of the radio astronomy community as those at all the other big instruments. Cooperation with other radio astronomy institutions, astronomers (particularly those at Leiden), and with the local community, was important. In many ways, the radio astronomers at Cambridge epitomized the new radio astronomy community, pushing interdisciplinary work further by getting the Cambridge University Mathematical Laboratory to computerize the synthesis of data.[67] In addition, though Ryle secured funding for his radio observatory from an industrial concern, he also secured the most overt statement of support for fundamental science untied from practical outcomes. At the opening of Cambridge's new radio observatory, the speech of Mr. Mullard of the Mullard Tube Company appealed to notions of pure science, largely along the lines of Appleton's address "Science for Its Own Sake:

I am only an industrialist, and we are supposed to be a hard-headed and practical breed—and in many ways we are. But that doesn't exempt us from a broader understanding which is not always based on enlightened self interest. People sometimes ask me what practical outcome I expect for my company from this Radio Astronomy venture and I have to confess that I just don't know, which is probably a very good thing because otherwise I might be accused of having ulterior motives. I don't suppose Lord Rutherford had much idea of nuclear reactors when many years ago he and his colleagues were unraveling the mysteries of atomic structure in the Cavendish Laboratory. One doesn't have to see the end right from the beginning and some things are worth doing for their own sake alone.[68]

Mullard hadn't been drawn to radio astronomy by practical motives, and he confessed to the audience that he didn't know what his company could expect from the sponsorship of a radio observatory. He seemed to have adopted Appleton's idea that practical results were not a necessary condition for supporting or pursuing work in a scientific field.[69] Ryle did offer Mullard the potential to solve the debate between the rival cosmological theories, the steady-state theory and the big-bang theory. As Ryle told his new patron, Cambridge's "present work has already produced strong evidence in favour of the idea that the universe is evolving." "One of the lines on which we hope to be able to proceed . . . [is] to distinguish between the various theories of the origin of the Universe. . . . This [and] other investigations which, though less exciting, seem equally important."[70] The result might be exciting, and the radio observatory's telescope unique, but the scientific style remained firmly communal. By offering cosmology, Ryle and the Cambridge group pushed the interdisciplinary boundaries of radio astronomy once more by appealing to cosmologists—themselves a new interdisciplinary community—for cooperation.

Ryle evinced his community-style science through interdisciplinary and international cooperation, the ideals of science without practical constraints, but also through
Cambridge's emphasis on pedagogy. Indeed, much of the strength of Ryle's case for
a giant instrument at Cambridge was attributable to the expansion of the Cambridge
group. In 1954, Ryle's Cambridge group had been strengthened with the addition of
eight graduate students; by 1956 the total number of staff members, assistants, and
students was 20.[71] Students became more adept astronomers as well as technically
proficient radio astronomers, and looked to expand their knowledge of astronomical
problems and radio applications. For example, one of Ryle's students, Conway, was
eager to spend a year after completing his PhD work at a center "preferably where
there is also optical work," and evidently Leiden was his first choice.[72] Ryle's request
that Jan Oort and the Leiden Observatory accept Conway came after Oort asked for
polarization measurements of radio emissions from the Crab Nebula. Exchange, cooperation, international, and interdisciplinary—this was the radio astronomy community
in action.

Leiden Observatory was a popular destination for students of radio astronomy. Its
director, Jan Oort, was receptive to the idea because, although Leiden had world-class
astronomers, they had to farm the construction of the radio telescopes out to engineers. "Taffy" Bowen and Joe Pawsey had sent an Australian, Eric Hill, to Leiden. In
1956, in the midst of the parabola-interferometer crisis, Pawsey expected the return
of Hill, who, he hoped, would bring with him a tacit knowledge of astronomical
practice. Unfortunately, Hill seems to have suffered some sort of breakdown about
two weeks before defending his thesis. He abruptly canceled his promotion ceremony,
and there was nothing that Oort, van de Hulst, or Maartin Schmidt could say to
change his mind. In Leiden, Oort was quite distressed. A few years later, Hill appeared
to be making progress in writing up his results for publication, but late in 1957
he still hadn't finished and both Oort and Pawsey feared he never would.[73] Not
all community-building endeavors work out, but the mechanisms are evident: exchanges of people result in the placement of students and secure linkages between
various practitioners groups with the expectation of a trade in expertise between
multiple disciplines, instruments, and methodologies. Though Eric Hill never did
graduate from Leiden, he did nonetheless strengthen the bonds between the Australian
radio astronomers and the Dutch astronomers, each of whom felt partially responsible
for the result.

Bernard Lovell, at Manchester, also sought cooperative student exchanges with
Leiden. Even short visits potentially offered great benefits for both sides. "Jennison
and R. D. Davies," Lovell wrote to Oort, "would very much like to come to Leiden

for a few days in order to see your hydrogen line work. I hope that it will be possible to arrange this because they are responsible for building the hydrogen line equipment which we hope to use with our new telescope, and it would be most useful if they could have some detailed talks with your own engineers. They also have a lot of new results from our present equipment, and would like to talk about these as well."[74] Significantly, Oort examined Davies' PhD thesis several months after this visit and was critical of Davies' overreaching attempt to do too much in the way of radio and optical studies. Oort commented that the thesis would have been better if Davies had concentrated on one or the other.

In short, by 1960 even the local education of graduate students in radio astronomy had been shaped to meet the intellectual and social demands of the new community. The earliest participants may have lacked formal training, but the patterns of community-style pedagogy were firmly entrenched. Indeed, by the 1970s, graduate students at Cambridge, Ryle insisted, got "six or seven lectures a week on astronomy during their first term as a very necessary grounding to their future work."[75] By then, the process of community formation had successfully integrated radio techniques into astronomy and had fully transformed radio physicists into radio astronomers working at radio observatories. The establishment of the communities' landmarks, the formation of networks, and the act of training "radio astronomers" had blurred the very distinctions that had made formal courses necessary in the first place.

THE REVOLT OF THE PHYSICISTS

The centrality and near ubiquity of parabolic type radio telescopes as the standard instrument of "radio astronomy" by the early 1960s was never a foregone conclusion. Lovell at Jodrell Bank, then Bowen at Parkes, championed the concept of a parabolic-style radio telescope as an instrument of a recognizable form to astronomers. Meanwhile, as we have seen, interferometer-type radio telescopes remained very useful in resolving radio sources and in making identifications that were closer to the expectations of the optical astronomers. In the process of community formation, Martin Ryle's commitment to interferometer-style radio telescopes didn't isolate him from his fellow radio astronomers or from the broader astronomical community. His young protégés cooperated with Caltech's optical astronomers around the same time that Edward Bowen's protégés did so. Thus, regardless of any group's commitment to any particular design of radio telescope, parabolic or interferometric, the more significant decision remained the extent to which the instrument fully integrated a sense of the international and interdisciplinary community.

The real challengers to the radio astronomers were alternative forms of social and intellectual community, not alternative instrument designs. That challenge came most vocally from a small subgroup within Australia's Radiophysics Laboratory that, in the mid 1950s, launched a major revolt against the parabola as the standard form for a large radio telescope, and sought to realign the radio astronomers with the physicists. Bernard Mills and Norman Christiansen, who advocated large interferometers of local design, led the revolt. Mills and Christiansen took advantage of the funding crisis of 1955, which temporarily halted construction of the Australian radio telescope at Parkes, to argue that the interferometer, not a parabola, was the correct form for a giant radio telescope. They claimed that "cross-type" interferometric systems were cheaper and were better at resolving radio sources. Like Ryle's, Mills' cross interferometer possessed two extended aerials crossing perpendicular to one another, producing a high-resolution instrument by combining the interferometric pattern from the two arms. Late in life, Mills remained convinced that "it was quite obvious that the limiting factor in doing a survey of the sky and making measurements on radio sources was the resolution of the instrument rather than its sensitivity. It was easy to see sources but they all ran together and you couldn't resolve them out. It was thinking over problems like this which led me to the thought that one could improve the resolution at the expense of the sensitivity by constructing a cross-type radio telescope."[76]

John Bolton's move to Caltech tipped a delicate balance within the Australian radio astronomy group. His departure deprived Edward Bowen of a stalwart ally and opened one of the major battles within the radio astronomy community. When the revolt was over, Mills and Christiansen left the Radiophysics Laboratory to construct a large interferometer under the patronage of the maverick physicist Harry Messel in the University of Sydney's physics department. Afterward, Christiansen spent nearly two years in the Netherlands in the early 1960s building the Benelux Cross radio telescope, which Jan Oort had been planning for several years. When Pawsey died suddenly in 1961, Mills and Christiansen emphasized Pawsey's commitment to "the string and sealing wax tradition" of the Cavendish tradition in their obituary.[77] Both Mills and Christiansen maintained for the remainder of their lives that the large radio telescope at Parkes had fundamentally undermined Australian radio astronomy.[78]

In a context of a struggle about the proper form of community, one can appreciate that Mills and Christiansen's almost lifelong antipathy toward the Parkes radio telescope stemmed from their alternative vision of radio astronomy in line with the Cambridge style of technically innovative but highly localized and isolated science. The pair certainly played their hand strongly and well. With only part of the funding available in 1955 for a projected giant radio telescope project, a working group within the Radiophysics Laboratory gathered to consider the laboratory's options. Though having secured a substantial portion of the necessary funding, Edward Bowen seems to have already been a limited voice in these discussions, adamantly supporting the single giant parabola despite clearly not having the financial resources to begin construction. Christiansen and Mills seized the opportunity to push their proposal for a giant interferometer that could be built with the money already donated by the Carnegie Foundation. The Radiophysics Laboratory group, they concluded, should consider "crosses capable of very high resolution." A cross-type interferometric radio instrument would still detect a considerable number of sources, they suggested: "a 5,000' cross, with a resolution of 1', should detect about 1,000 sources at 1420 Mc/s [i.e., 21 centimeters—the hydrogen line]." Christiansen thought that in solar studies "a large paraboloid would hardly give the minimum resolution needed for useful observations," and added that "such resolution can better be obtained with interference system." Though Mills conceded that the giant parabola would be more sensitive than any interference system, the cross interferometer would be easier to construct and would have a distinctly smaller price tag. Additional support for crosses over parabolas came from the Radiophysics Laboratory's own "radio telescope planning committee," which gave the familiar argument of Australia's peculiar geographical position a new twist. Australia's geographical isolation meant that the country didn't require, as yet, a high-resolution instrument with a wide frequency range. "It was agreed that any large aerial should take advantage of Australia's geographical position, particularly in view of future U.S. competition."[79] Thus the southern hemisphere's only radio astronomy group needed a broad survey instrument, and would await improvements in equipment and technique before moving into detailed investigation of identified sources.

Independently, Mills and Christiansen managed to gain much of the funding for their alternative instrument. Yet dreams of an interferometer with arms perhaps 3 miles long evaporated because there was no possibility of building both a giant interferometer and a giant parabola. As Mills and Christiansen continued to argue that substantial results could still be obtained from simple and inexpensive equipment, Bowen took to attacking them with increasing vehemence. For Bowen, it seemed that Mills and Christiansen characterized radio astronomy as "string, a few wires up in a field, and, lo and behold, a miraculous result!" Defending the giant parabola, he asserted that "nothing could be further from the truth." Even the minor interferometer-style radio instrument that the Australian group did build, the so-called Chris-Cross, "will have taken three years" to build, and cost 90,000 pounds, all "before any research results have been achieved."[80] Bowen continually emphasized that the only way to ensure that the Australian group could maintain its influential position within the radio astronomy community was to commit wholeheartedly to the giant parabolic radio telescope. Pawsey recognized that line radiation studies, like hydrogen's, desperately required "much higher resolution," which would only come from "200 to 300 ft. apertures."[81] Many within the Radiophysics Laboratory remained unconvinced that the single large parabola represented the end point of radio telescope design.

The arguments cited above were only the overt arguments between technological alternatives. As is true of much of the history of science, the debate about price, resolution, and sensitivity concerned the rapidly changing nature of science itself. When Mills and Christiansen advocated waiting for developments in technique, they understood radio astronomy as continuing to change. In contrast, Bowen accepted the single parabola as an established technique and viewed the radio astronomy community as approaching stability, certainly enough to commit more than half a million pounds to a single instrument. In addition, much of the tension between parabolas and interferometers stemmed from competing views about the balance between astronomical and radio-physical expertise that a radio astronomer should have. In the end, although Bowen's deputy Pawsey supported the continued development of interferometers, his commitment to Bowen's parabola reflected his view of the weakness of the Radiophysics Lab's astronomical credentials. He complained to Jan Oort: "Kerr, Mills, Bracewell, Christiansen and the rest are good physicists but their astronomy is scrappy."[82]

AN AUSTRALIAN IN CALIFORNIA

As a storm erupted over the choice between parabolas and interferometers, or more exactly between laboratory and observatory practice, a penultimate component of the radio astronomy community took shape. Edward Bowen's protégé John Bolton arrived in California in February of 1955 charged with building a radio telescope for the world's center of optical astronomy. Lee DuBridge, always the optimist, hoped that he and Bowen might yet "build our two big dishes together."[83] Of course, Caltech had been involved in building the radio astronomy community for many years before their own radio observatory took shape under Bolton's leadership. Courses had been run for several years already at Caltech, and the astronomers from the combined observatories (Mount Wilson and Palomar Mountain) had provided valuable assistance to the various young visiting radio astronomers from Australia and Britain. But the appointment of Bolton, who was as close to a professional radio astronomer as could be found in the mid 1950s, signaled the stabilization of radio, optical, and indeed all-wave astronomy at Caltech.

News of Caltech's new instrument program must have spread widely and rapidly. The first submission of a radio telescope design for Caltech—evidently unsolicited— came from the Fuller Research Foundation. The Fuller Research Foundation, established by R. Buckminster Fuller to espouse geodesic design, eagerly contributed proposals and scale models of possible designs both to Caltech and to Australia's Radiophysics Laboratory. Edward Bowen and Jesse Greenstein examined the Fuller proposal with "a good deal of skepticism." "The aerodynamic aspect seems to be quite unfavorable," Greenstein noted dryly. "Most people think that in a decent wind the thing will blow away down the valley."[84]

As designs appeared, extracting nearly US$90,000 from the Office of Naval Research to expand Caltech radio astronomy became Bolton's first task in California in 1955.[85] Caltech promised to initially support radio astronomy for expendable "electronic equipment" and only Bolton himself and a two-year fixed-term Senior Research Fellow. In the wake of the formation of the National Science Foundation, the ONR countered the more respectable source of funding for science with ever-larger offers of large sums of undirected money. Under Bolton and DuBridge, Caltech radio astronomy pursued patronage from the military-industrial complex more vigorously than any other program, and more successfully. One day, for example, an ONR representative showed up at Greenstein's door and asked "Do you want a hundred thousand dollars?" Interviewing Greenstein in later life, the historian Spencer Weart asked "This guy had no idea what you might do with the hundred thousand dollars?" Greenstein replied: "Not in detail. He just said he wanted to support science—I guess, a prestige science, astronomy, and a big observatory." And although Greenstein openly admitted that he would "take money from bombardiers or anybody for astronomy,"[86] in fact the incident reveals a subtle change in what the ONR was supporting scientifically. It had moved away from individual scientists and their projects and toward an emphasis on large instruments. Lee DuBridge hoped to take advantage of that shift, but Robert Bacher worried that doing so might leave the Caltech program with a large and expensive centerpiece instrument but no money for staff, research programs, or future electronics development.[87]

Bolton's larger vision for Caltech's radio observatory, which was true to the radical social transformation that community-style science required, took DuBridge by surprise. DuBridge assumed that Bolton would arrive in Pasadena full of ideas about cross-style interferometers or maybe big dishes. Instead, DuBridge told Merle Tuve of the Carnegie Institution, Bolton had "evolved an idea for using two . . . dishes . . .

FIGURE 4.4
A design for a radio telescope proposed by the Fuller Research Foundation, May 27, 1954 (Australian Archives, series C3830, A1/3/11/1, part 2). Reproduced courtesy of Australian Archives.

FIGURE 4.5
John Bolton (far left) with workmen in front of one of the two 90-foot dishes of Caltech's future Owens Valley Radio Observatory (California Institute of Technology Archives, OVRO 1.15–1). Reproduced courtesy of California Institute of Technology Archives.

one fixed and one movable."[88] The two dishes, Bolton claimed, would give the "sensitivity and resolving power approaching that of the big crosses now under construction," with the added benefit of not limiting Caltech's program to specific wavelengths or fixed regions of the sky.[89] According to a press report issued the following year, the twin parabolic telescopes' "exceptionally high resolution" would enable astronomers to "distinguish between objects that are close together," allowing "further identification of radio sources with visible objects."[90] Caltech's radio telescopes would be steerable "in order to detect very weak signals, in particular, for work on the hydrogen line," and would be paraboloid "to operate over a wide range of frequencies" yet also interferometric to ensure sufficient "resolving power for the position and angular size measurements."[91] Bolton's vision was of two steerable 80-foot radio telescopes whose goal would be "the identification of some 500 radio stars." Bolton's imposed limit on the sheer size of each dish acknowledged that the single-minded search for ever-fainter sources wasn't really Caltech's concern. Atop the mountains, Minkowski and Baade remained primarily interested in finding radio objects that had optical correlates.

Under Bolton's guidance, Caltech's radio observatory took on its distinct shape: two parabolic dishes joined to form an interferometer. The Owens Valley Observatory, as it became known, continued international and interdisciplinary cooperation, completed the transformation of radio practice into astronomical practice, and, once more, reaffirmed pedagogy in radio astronomy. Warren Weaver of the Rockefeller Foundation easily envisioned the unity of radio and optical astronomy. Bolton's new equipment, Weaver predicted, "will supplement the outstanding equipment they [Caltech] have in the optical field."[92] The two dishes would be on a rail track, so that they could be used either individually or in combination. This "versatile system," with a variable axis and a wide frequency range, would be "unique." The "frequency coverage of the equipment and program," Bolton noted, "has been chosen as 400–1600 Mc/s. Relatively few observations have been made in this range and there thus exists the possibility of detecting new and unsuspecting phenomena."[93] The power of community again, gaining global unity while permitting local differences.

Bolton claimed that only 50 of the 2,000 radio-identified stars had optical correlates. And of those 50, about 10 were "grand catastrophes" whose radio emission was not well understood. Caltech would not continue the search for ever-greater numbers of radio sources, as "most workers" had, but would turn its attention to "the more difficult task of identification."[94] Astronomy, Bolton remarked to Greenstein, had at last reached the stage "where the conventional astronomer would like to have some say in what the radio astronomer should do instead of merely turning the telescopes in his wake."[95] For Greenstein and DuBridge, all-wave astronomy turned the astronomers' world upside down. Optical astronomers became "conventional," while radio took priority in dictating research programs, and telescopes turned at the whims of radio astronomers.

Even with the cheaper, smaller dishes, the estimated cost approached US$450,000. Bolton's two-element approach traded sensitivity for resolution and frequency breadth, but in reality the telescope didn't require radically new developments in instrument form, merely in application. "The individual antenna and their costs are similar to other proposed steerable reflectors in the United States and Holland," Bolton explained to the ONR; "however the increase in scientific value of the two element system is very much greater than the increase in cost."[96] The moral economy of community-style science sought results at a rate faster than the rate of increasing cost. Among Bolton's many supporters, Tuve liked the idea of taking an established dish design and employing it in a novel way. Tuve pushed DuBridge to encourage Bolton to initially build the planned 84-foot dishes, but believed the leap to 120 feet would be "a straightforward design problem" for the engineers. Bolton recruited the Kennedy Company of engineers, who had already designed and built Harvard's 60-foot dish and who routinely built smaller dishes for military radar. Doubting that the Carnegie Institution's radio astronomy group would ever expand its instrumental facilities, Tuve thought Caltech was in an ideal position to fulfill an intermediary position between small and giant dishes.[97]

As the technical vision of the Owens Valley radio observatory took shape, the new radio astronomy community was formalized within Caltech. Ira Bowen and Jesse Greenstein's preliminary work on the ONR proposal incorporated requests for a senior astronomer, a senior electrical engineer, a junior physicist, and a senior technical assistant.[98] The equality of the level of the astronomer and the electrical engineer, with the physicist relegated to a junior position, stands out here. In fact, Greenstein found, much to his surprise, that astronomy appointments "had the backing of the physics group from the beginning, . . . then, oddly enough, electricity, and magnetism and electrical engineering, because of radio astronomy."[99] The process worked in both directions. An astronomy appointment in a field like radio astronomy necessitated that students wander over and take courses in electrical engineering, thereby strengthening both groups' student numbers. This fluid crossing of disciplinary boundaries stemmed from the realization that "there wasn't a single person in radioastronomy you could appoint a professor at Caltech."[100]

Until Bolton's position at Caltech became permanent, Greenstein and Bacher had relied on a plentiful supply of temporary visitors to teach radio astronomy. In the year before Bolton arrived, Otto Struve, at Berkeley, proposed that Ronald Bracewell, an Australian radio astronomer who was slated to visit Berkeley the next year, give a series of lectures at Caltech. Struve was bringing Bracewell to Berkeley to "give a course in radio astronomy" along with "a seminar for graduate students." Bracewell's position as an instructor in radio astronomy was heightened by the publication of his *Textbook on Radio Astronomy*, co-authored with Joe Pawsey.[101] But Greenstein couldn't schedule Bracewell for anything other than "one lecture" during the "regular colloquium series." Caltech, it seemed, had had "a rather great surplus of visitors in the field of radio astronomy," and the coffers for guest speakers were empty.[102] Caltech and the mountain observatories already had hosted F. Graham Smith from the

Cavendish and Bernard Mills from Australia's Radiophysics Laboratory, and more offers came in every day: Edward Bowen mentioned to DuBridge that he had several people with "itchy feet" who hoped "obtain positions in the USA." Bowen offered Caltech Jim Roberts, who, though "less practically inclined" than Bolton, was "very strong on the theoretical side."[103]

Radio astronomy advanced rapidly in California. By April of 1956, Caltech's 32-foot dish—built within sight of the great Palomar telescope—was in operation. Handled by experienced radio astronomers, the new telescope required only about a week's tuning before Bolton's new assistant, Gordon Stanley, happily heard solar noise at eight times receiver noise. Unfortunately, the beam width appeared to be about 20 percent "less than theoretical"—Bolton wasn't sure how this had happened, but he suspected an underfed center and extremity of the dish.[104] Still, Greenstein—reassured by the early experience with the 32-foot dish—expressed confidence that Bolton and Stanley could "well handle the electronics and scientific engineering, and that if we have really good mechanical engineering, the bigger construction will go very well indeed."[105] The proximity of the radio and optical telescopes "naturally," said Caltech's news release, allowed "a co-operative program among the optical and radio astronomers." In fact, the 32-foot telescope had almost no research potential. Rather, it served "mainly as a pilot model"; its actual role, like that of Harvard's early radio telescope, was "as a training instrument for astronomers and electronics personnel."[106]

The instrument worked in unusual ways to create cohesion between the electronics engineers and the astronomers at Caltech. With one or two members of the radio astronomy group staying atop Palomar Mountain most days after April of 1956, the radio and optical astronomers moved, literally, into the same spaces. The "Monastery," the traditional housing for researchers, soon overflowed with guests.[107]

Caltech's adventure into radio astronomy, unlike Harvard's, emphasized significant research programs alongside the training of disciples. Bolton's new colleagues soon discovered that he avoided participating in formal teaching programs. Greenstein recalled that Bolton "wouldn't teach. . . . He probably couldn't, to be exact. . . . It was just that [he] was the non-teaching type . . . dead set on building up instrument excellence and new kinds of antennae; [he remained] awfully remote from the academic or even the interpreter's side."[108] While Bolton evaded his teaching responsibilities, Greenstein roped more Caltech visitors into teaching courses in radio astronomy. Kevin Westfold, another visitor from Australia's Radiophysics Laboratory, stayed at Caltech during 1957 and 1958, ostensibly to give a single course in radio astronomy. Since he subsequently remained at Caltech, working for Bolton during 1959, Greenstein hoped that he could be encouraged to teach the course again, since there were now six graduates students in "Astronomy, or Radio Astronomy, who have not had this course." Greenstein was "anxious" to keep a course on radio astronomy present in the course catalog for most years, hoping "to build up interest in Radio Astronomy as quickly as possible among the students in Astronomy and in other fields, notably Physics and Electrical Engineering."[109] Westfold's radio astronomy course again presented radio and optical astronomy as essentially complementary endeavors. After the introduction, "observational methods" were discussed before "thermal noise" and "radioemission by the moon." "Optical investigations of the sun" preceded "radioemission from the quiet [and active] sun."[110]

By mid 1957, Bolton had the "21 cm work . . . going very nicely" at the Owens Valley Observatory. Back at Caltech, Greenstein happily reported to Tuve, "we have a few graduate students in radio astronomy, so I think things will move along pretty well."[111] Caltech's integrated approach of novel instrumentation, pedagogy, and community paid dividends for years to come. Decades later, one of Bolton's students, Ken Kellermann (PhD 1963) thought that Bolton may have been the only Caltech faculty member to avoid teaching duties, but believed that Bolton had produced an excellent cadre of students nonetheless. Robert Wilson, a Nobel laureate for his discovery of the microwave background, was one of Bolton's first students. Barry Clark designed the Very Large Array in New Mexico after gaining experience at Owens Valley. Other students of Bolton ended up directing radio observatories in Australia, in the United States, in Europe, and in India.[112]

As Jesse Greenstein wrote to Maarten Schmidt, the newest appointment at Caltech from Leiden University's Observatory, radio astronomy "may be a field which one should consider very seriously for the long term."[113] By 1960, only a few years after Sputnik, the cooperative, international community of the radio astronomers realized the benefits of productive and effective patterns of work. Atop Palomar Mountain, Rudolph Minkowski and Walter Baade continued to receive more accurate positions of the radio objects from the radio astronomy groups in Britain and Australia and, increasingly, in the United States. That same year, Minkowski, utilizing radio positions from Martin Ryle's new interferometer at Cambridge and from John Bolton's two-element parabolic array at Owens Valley, made an optical identification of one of the sources on Robert Hanbury Brown's list. A year later, Allan Sandage at Palomar Mountain photographed another radio source from a more accurate radio position supplied by Thomas Matthews, one of John Bolton's early students at Caltech, who now was also at Owens Valley. That object, 3C48, would open a new research window for radio astronomy and cosmology. Called a Quasar, it helped elucidate the structure of the universe at vast distances.[114]

The Jodrell Bank radio telescope first moved in June of 1957 and was completed late in August. It had taken five years to build, and in that time about half a dozen major radio telescopes had been planned and funded. During those five years, the purpose of Jodrell Bank had changed so much that by the time the telescope began listening to the universe its surface already required a substantial upgrade. Lovell also sought to upgrade Jodrell's Bank's staff. He wanted to support the ongoing research projects of two near-PhDs from Manchester and to appoint another member of the staff to a fellowship. Lovell's accountancy permitted at least "two senior fellowship posts" to be filled, and he hoped to attract "some really senior workers from abroad to work with us on the large telescope."[115]

Lovell's student fellowship appointments indicated the new identity of Jodrell Bank. Both were physics graduates from Manchester who had already undertaken work at Jodrell Bank. J. V. "Stan" Evans was especially concerned about the electron content of the ionosphere, but also contributed to Lovell's project involving radar echoes from the moon. Lovell later mentioned to the sociologists Edge and Mulkay that he had attempted to entice Evans back into radio astronomy, but Evans had ended up as a lecturer in the Engineering Department at Cambridge.[116] However, the other student fellow, Barry Rowson, became a radio astronomer when he used a radio interferometer to measure the size of galactic radio sources, very much along the lines of Bolton at Caltech, Ryle at Cambridge, and Mills and Christiansen in Sydney. Both Evans and Rowson published in the *Proceedings of the Physical Society*, whereas nearly all of the earlier surveys of the radio sky funded by the Office of Naval Research had been submitted to *Monthly Notices of the Royal Astronomical Society*.[117] On such publication metrics, Lovell's telescope was now distinctly an instrument of physics. Measuring disciplinary affinity by publication outlet instead of training reveals one of the sociological fallacies of Edge and Mulkay's study. Rowson and Evans came of age in a physics department, and it is important to contextualize Lovell's endeavor via his departmental situation. Evans repaid Lovell, the Leverhulme Trust, and his deferral from the National Service by tracking Russian satellites later that year.

Lovell and Jodrell Bank's best ally showed up only three months after the dish first moved. It was Sputnik that got Jodrell Bank "onto the front pages of the world's newspapers."[118] And in 1960, the United States launched Pioneer V, a probe carrying an array of scientific instrumentation into deep space. Jodrell Bank and NASA became institutionally linked during the Pioneer mission. In the short term, that affiliation helped pay off the outstanding debt on the telescope; in the longer term, it tied a radio telescope to the space program in the scientific, political, and public arenas.[119]

5 SIZE

When you begin to think of using a national facility—we had no idea of what was involved yet, but that it would be big. Like an accelerator.[1]

—Jesse Greenstein, Caltech astronomer

By June of 1956, the plans for the new National Radio Astronomy Observatory (NRAO) were all but finalized. After nearly three years of negotiation and review, the National Science Foundation, a consortium of universities represented by Associated Universities, Inc. (AUI), and interested physicists and astronomers concluded that by 1961 a radio telescope 140 feet in diameter would be built at Green Bank, West Virginia. The centerpiece instrument of a national facility for radio astronomy would eventually, the panel reported, bring outstanding research opportunities to American radio astronomers—opportunities that would rival and perhaps even surpass those offered by the giant radio telescopes of Britain and Australia.

Only one substantial question remained outstanding before American astronomers could expect construction to begin: Exactly who would use the facility? Merle Tuve, leader of the Carnegie Institution of Washington's radio astronomy project and, moreover, head of the National Science Foundation's Panel on Radio Astronomy (which was charged with the review of the entire NRAO project), claimed that he could identify no one who would rather use a national facility than use local resources. Writing to the rest of the NSF Panel, Tuve argued as follows:

Harvard has Heeschen and Matthews and several students. These young men have just acquired a 60 ft equipment, which is quite a lot for them for a while yet. NRL [the Naval Research Laboratory] has McClain, Lilley, Roman and several others; they have a 50 foot and a new 84 foot equipment not even yet set up. Michigan has Haddock and one advanced engineering student; they are erecting a 28 foot and a 60 foot equipment. Cornell has Marshall Cohen and several equipments. We here have five or six working in radio astronomy, and several modest equipments; but you may be sure that the NSF Green Bank installation as guided by the AUI is not being planned by or for our Carnegie group. Ohio State has Kraus and several students, but they are busy with their large helix installation and are building a large tilting reflector for which we supplied funds. They are not much interested in the steerable parabolas nor in the AUI plans. Cal Tech has over half a million dollars and two 90 foot dishes under construction, plus others. Stanford has Bracewell and one or two students; they have funds approaching \$200,000 from the Air Force for an installation yet to be built. . . . Among those who hope some day to do something are Rensselaer, Case, Yale, Illinois and Penn, and among those who sympathize with the idea of having radio astronomy happen some place and can give it an occasional technical or administrative boost are Berkner, Menzel, Townes, Dicke, Wiesner, Goldberg, Gordon, Seffert and perhaps others. Where among these are the sound research men who need this new five million dollar facility which we have urged? Shall we now decide we must import them from Manchester or Sydney? Our men in the universities are all building big equipment on funds supplied by ONR, OOR, NSF or the Air Force. Can we name even one first-rate man who is prepared to accept personal responsibility to make this added "National Facility" a wise and fruitful venture for the NSF?"[2]

As we now know, the physicists gained much from a sheer abundance of people. In contrast, as Tuve points out in his almost complete listing of the American radio astronomy community, a "national facility" for radio astronomy had no comparable wealth of expertise. The radio astronomers remained thinly spread across a dozen institutions, and the community consisted largely of graduate students. Here was the emerging contrast between community-style science and complex-style science, which was most noticeable when it came to funding large scientific instruments. The National Science Foundation should not be preoccupied, Tuve argued, with the path taken by the Office of Naval Research and the Air Force. The NSF should avoid supplying big equipment and emphasize instead programs of research that men might bring to a scientific facility.

By the late years of the twentieth century, big-equipment projects for astronomy such as the Gemini program, the Hubble Space telescope, the Very Large Array, and the Very Long Baseline Interferometer—what we now call "big science"—had revealed wondrous visions of the universe but had also concentrated material resources to an unprecedented degree. The new reality of "big science" was that an astronomer became a "small part of a very big team," as the director of the Gemini program, Matt Mountain, told the historian Patrick McCray.[3] Larger facilities distanced astronomers from direct contact with their instrument, forced astronomers into large and amorphous teams of international collaborators working on common research problems, and necessitated the pursuit of bigger budgets and larger facilities and instruments. But it is the steady loss of lone researchers that has most deeply troubled modern astronomers, and so they have grown fearful of the social costs of giant tele-scope projects even while they have reveled in their new vision of the universe.

Big science, as many authors have pointed out, grew from the increasingly intimate relationship between science and the nation-state. It was no wonder, then, that the issue of importing foreign expertise into a supposedly national facility became grounds for challenging the entire project. Tuve's remarks were squarely aimed at Associated Universities, Inc., and especially its chairman, Lloyd Berkner. After AUI's successful bid to build the Brookhaven National Laboratory for nuclear physics, Tuve regarded Berkner's push for a national radio astronomy facility as little more than an exercise in scientific parochialism. As Tuve noted, the centers of radio astronomy were in Australia, Britain, and the Netherlands.

Tuve advocated radio astronomy as a community-style science. In contrast, Berkner envisioned the NRAO as emulating the dominant model of scientific practice, nuclear physics. He saw radio astronomy adopting the complex style of nuclear physics. For Berkner and AUI, the NRAO, like Brookhaven, would be a large centerpiece facility through which practitioners would flow from other institutions. The centers of radio astronomy in the United States were far smaller than either their Australian or their British cousins and were mostly based at universities. Moreover, most of the new field's practitioners were still graduate students, advised by men from a diverse array of disciplinary cultures. John Kraus at Ohio State was an electrical engineer, Marshall Cohen at Cornell a physicist, Bart Bok at Harvard a galactic astronomer, and John Bolton at Caltech a former radar operator.

As the chief power broker behind the push for a national instrument for radio astronomy, Berkner attempted to shift the communitarian culture of the emergent science toward the complex style that Brookhaven embodied.[4] The example of Brookhaven, Berkner explained, "provides able scientists the opportunity to carry on the most advanced research requiring great and expensive facilities, without loading the academic staff of any single university with the over-burdening task of utilizing such an expensive facility to the capacity that its cost requires."[5] Thus, the advantage of a national facility was abundantly clear to Berkner and the New England universi-ties that first proposed a national radio astronomy observatory. Only the federal gov-ernment could sponsor new instruments as large as those that nuclear physicists and radio astronomers were demanding. As construction costs alone exceeded the capacity of any single institution, large-instrument programs became appealing as ways to employ more physicists and astronomers, hence enlarging both communities.

Such cultural battles over styles of science repeated the struggles the British group under Bernard Lovell and the Australian group under Edward Bowen had gone through to acquire large instruments, and the social costs they had incurred. As the last chapter illuminated, the lackluster response of the military-industrial complex in either Australia or Britain saw the radio astronomers advocate their work as even more communitarian and directed toward higher goals. Yet within the British government the model of national scientific facilities had considerable appeal. Sir Ben Lockspeiser, secretary of the Department of Scientific and Industrial Research, the organization responsible for much of the funding for Bernard Lovell's dish at Jodrell Bank, noted that, although "the capital cost is considerable, and puts the equipment into the same class [as] some of the new cyclotrons," there "may be much to be said in favour of having it at a central institution where it is available for all who wish to use it rather than at a particular university. In fact Brookhaven was founded for this purpose."[6] The social organization of nuclear science implied centralized control and large state-sponsored instruments and facilities, and offered a standard model for the operation of science throughout the Cold War. About the only thing that stopped British science from copying American science, John Krige argued, was that British universities, though empowered with vast new funds and equipment to pursue nuclear research, didn't possess "centralized military-type control" and didn't suffer "the organization and coordination [of research] on an industrial level."[7] In other words, they didn't possess the interlocking networks of the American complex, and thus they couldn't fully adopt complex-style science.

In the United States, by contrast, the existence of parallel models of giant scientific facilities, not to mention the considerable talents of the technocrat Berkner, presented some American radio astronomers with their best opportunity to build powerful new telescopes with powerful new patronage. Through the comparison, we can see how Tuve's resistance to Berkner was a microcosm of the broader challenge the radio astronomers presented to the wider scientific community, most especially the physicists. The radio astronomers also seem to complicate the story of Cold War–era "big science"—a story in which American nuclear physics is seen as the standard model.[8] The age of "big science" saw dramatic increases in funding, often through the branches of the military. Such funding usually resulted in concentration of resources, growth of specialized workforces, and attachment of political significance to scientific projects.[9] But all participants recognized the cultural battle they were fighting even as they built their vast technoscientific facilities. For the Brookhaven Lab's director, Alvin Weinberg, for example, "big science" was something more than monetary. Weinberg appealed to a higher calling. "Big science," he said, displayed an ability to embark on great projects for seemingly higher purposes that might legitimize a nation's greatness. Accelerators were directly comparable to the pyramids.[10] The idea of a higher calling for science also linked complex-style science to the ideal of pure science. Despite the obvious military uses of oceanographic surveys, larger accelerators, or radio telescopes (charts for submarines, studies of atomic particles, and listening to Soviet radio signals, respectively), Weinberg could still view them primarily as examples of "pure science."

Radio astronomy might easily have become enrolled into the complex; it was certainly courted. In 1953, for example, the Office of Naval Research re-released its announcement of "a broadly conceived program of basic research in astronomy and astrophysics" (originally published in the *Astronomical Journal*), sending it individually to every member of the American Astronomical Society. Because the creation of the NSF had caused the "re-evaluation of the Navy's responsibility in support of research" in astronomy,[11] the ONR looked to resume "support of basic research in astronomy with a modest program in those areas clearly relevant to the navy's scientific interests." Among those areas were astro-ballistics (the field of Fred Whipple's study of meteors at Harvard), upper atmospheric research, navigation, astro-gas dynamics, and radio astronomy. But, the announcement said, "any area of astronomical research, for instance, in which major advances or breakthroughs are foreshadowed cannot help but be of interest also to the Navy as the guardian of one of this country's outstanding establishments in the field."[12] In other words, the military-industrial complex sought potentially valuable technical expertise outside its already wide purview, and radio astronomy had grown large enough to be noticed. Yet Tuve's roll call of the field (reproduced above) supposedly highlighted the difficulty he saw in identifying anyone in the United States qualified to use a big dish, whereas for Berkner radio astronomy's parabolic dishes seemed to be straightforwardly expandable, as cyclotrons had been throughout the 1950s.

Tuve and Berkner, then, were on opposite sides of a community divided on the question of what role federal organizations should play in erecting facilities for communal use. In contrast with earlier battles over the measure and the meaning of radio astronomy, the form of the instrument was not the real issue. The real issue was whether local, cooperative, university-based programs should give way to centralized and federalized facilities.

One important aspect of any form of complex-style science was the dominance of corporate decision-making processes. Planning for the NRAO took place from 1954 until 1956, mostly in offices in Washington and New York. Significantly, the people making the decisions about the NRAO happened to be exactly those people who were already invested in local radio astronomy programs—among them Merle Tuve, Bart Bok, Jesse Greenstein, and John Hagan. They shaped the NRAO and many other smaller radio astronomy projects by applying to the NSF for funding. They then evaluated one another's research proposals—after politely asking the proposer to step out of the room.[13]

The creation of the NRAO was part of a larger cultural shift in science, especially American science, toward "national" facilities. But in the case of the NRAO the eventual form of the "national" radio telescope at Green Bank incorporated resilient notions of a dispersed scientific community, as we will see. The eventual telescope, 140 feet in diameter, was evidence of a substantial concession to smaller university-based radio astronomy programs. It also acknowledged the necessity of continued participation by practitioners from a wide range of disciplines. Finally, it surrendered the notion of "national" entirely. Perhaps Tuve's major achievement was the recognition that most senior researchers in radio astronomy were not even American. In terms of recruitment, American astronomers asked, where were the teams of expertise to effectively utilize a new facility? In 1954, when the initial planning documents were drawn up, David Heeschen was the only "trained" American radio astronomer. Was the "national" facility for him alone?

The National Radio Astronomy Observatory can be traced back to January 1954, when the Caltech astronomer Jesse Greenstein held a conference on radio astronomy that attracted an array of "astronomers, physicists and electronics men." Nascent plans emerged over the course of that weekend, but from the outset the question was whether radio astronomy and any new telescope would continue to have implications "not only in astronomy, but in physics, electronics, and electrical engineering."[14]

The 1954 meeting presented an opportunity to plan "a major laboratory" for radio astronomy.[15] Whether a radio astronomy facility was a laboratory or an observatory was still uncertain, but evidently there was general agreement that radio astronomy needed "a very large radio facility in the United States, sponsored by the federal government" and that "various universities [should] go ahead and expand or build radio astronomy facilities."[16]

Perhaps there was general agreement, but there remained considerable resistance to a centralized large facility. The astronomer Gerard Kuiper thought that "pouring money into gap fields can only lead to waste and disappointment."[17] Others saw the thin end of a wedge. Who would safeguard the interests of the entire astronomical community from the NSF's planning groups? Geoffrey Keller and Philip Keenan of the Perkins Observatory wrote to Greenstein to say that they were "disturbed by the implications of . . . a super planning group made up of a few of the most influential members of the Society." Astronomers, like many other Cold War–era scientists, were lured by the prospects of larger budgets and facilities but suspicious of any move toward science by committee. Keller and Keenan specifically questioned whether any super planning group would "necessarily be in the best interests of Astronomy as a whole in this country."[18] Greenstein himself was also cautious about removing radio astronomy from its university nursery too soon. Radio astronomy had risen to such a status that the federal government could realistically sponsor a national facility, but the work that justified the facility had been done by small groups based at universities, not all of them in the United States. In other words, Tuve and Greenstein seemed perplexed that, in the case of radio astronomy, the cooperative efforts of a number of small, independent, international groups of observatory researchers somehow warranted a centralized national facility. Greenstein cited the Harvard Observatory: "There seems to be little doubt that the Harvard project so far has been an extremely successful demonstration of the value of doing radio astronomy work in an astronomical observatory in conjunction with programs of that institution, in this case in the field of galactic structure."[19]

In a "Survey of the Potentialities of Cooperative Research in Radio Astronomy" that considered the possibility that several East Coast institutions might join together to confront the problems of research in radio astronomy, the Harvard astronomer Donald Menzel wrote: "Research tools are already complex and promise to become increasingly so," and "the cost of these tools is rapidly going beyond what a single institution can provide." Menzel gave his survey to Lloyd Berkner and AUI for consideration in March of 1954, and in April it came before the NSF's Panel on Radio Astronomy. The East Coast institutions saw an opportunity in radio astronomy because it required practitioners from several disciplines; indeed, the "cross-field nature of the research" was considered a boon to the new science of radio astronomy.[20] Menzel had succeeded Harlow Shapley as director of the Harvard College Observatory a few years earlier. As I have already argued, declining instruments and the development of local expertise brought from Harvard's physics department in the form of Harold Ewen influenced Harvard's entrée into radio astronomy. Like their West Coast compatriots, East Coast astronomers looked to radio astronomy to solve all manner of institutional, disciplinary, and instrumental problems.

The East Coast institutions emphasized site "accessibility" for their astronomers, as well as freedom from local radio interference. The Survey concluded that the easiest course might be the expansion of Harvard's own radio astronomy facilities, which had been active since 1952. Contiguous land was available. Members of the Harvard Observatory already dominated American training and research in radio astronomy.

"Competent graduate students" were an "important aspect of [Harvard's] program" for productive research in the field, and the institution, it was claimed, possessed the only academic program in radio astronomy in the United States.[21] When Jesse Greenstein (an alumnus of Harvard) initially evaluated the survey for the NSF, he commented that "it would be nice if the equipment could be put just on the property of the Harvard Observatory," but he doubted the proposal's wisdom in the long term, being especially "dubious about its being the ideal location for a big dish costing several million dollars."[22]

Support for a national facility came from some unexpected quarters, including small and isolated astronomical programs. At a conference on Astronomical Research in Small Academic Departments held only a few months after the Caltech meeting, there was a general consensus that a national facility would increase opportunities for the most instrumentally deficient researchers and teachers. Peter van de Kamp of the NSF summarized the participants' view: "A National Observatory could greatly improve the research opportunities for the teacher of astronomy in the small department. It could meet many of the problems brought out at the conference, especially if the research needs of the teachers of astronomy in small departments are specifically kept in mind in developing the program for the National Observatory."[23] Other astronomers not only supported federal sponsorship of astronomy but argued that the federal government's role in astronomy should be greatly expanded. Astronomy, said the Berkeley astronomer Otto Struve, needed special government support for its major projects: computing centers, radio astronomy, and a national optical observatory, all costing millions of dollars. Struve proposed three large computing centers—"not for astronomy alone, but for all basic sciences that are not materially aided by [the Atomic Energy Commission] or other governmental or industrial set-ups."[24]

Upon returning to their universities, the members of the cooperative survey committee reported on their initial meeting with Lloyd Berkner. At Harvard, Donald Menzel expressed enthusiasm for Berkner's ideas and derided the view, expressed by Jerome Wiesner of MIT, that a $2 million project would be "unwieldy." The Harvard College Observatory Committee discussed possible sources of funds for the project but tended to concentrate on the question of a site. "Accessibility" was "a primary consideration," and the site of the NRAO "should be near [Harvard] if possible," even though Harvard's Observatory Committee recognized that galactic studies would benefit from a southern location, perhaps in Florida, and saw clearly the political reality that advocating a site in Texas would almost certainly bring in the necessary $2 million. Harvard's astronomers remained adamantly against any work that would involve classified research.[25]

As Harvard concerned itself with a national facility's location, Jesse Greenstein of Caltech raised the important question of leadership. He admitted frankly that he did "not think there is anyone in sight in this country who could run the organization proposed by the A.U.I." Radio astronomers in Australia and Britain had come from radar backgrounds, but Greenstein doubted that Americans, even from "pure astronomical centers," would "be the same high-quality." Optical astronomers were conscious that newer and more glamourous areas of science were attracting the best potential researchers—people who had both skill in physics and electronics and an interest in astronomy. Radio astronomy needed to lure "bright young people," Greenstein insisted, like those who went to "work on atomic energy, nuclear physics, large accelerators, et cetera." Only then would American institutions have "the same bursting enthusiasm" as Australians or British ones.[26] In this sense, Merle Tuve found a powerful ally in Greenstein. Large facilities were useful only if there were large numbers of researchers and students to use them. "What we most badly need," Greenstein told his audience in 1954, "is people, rather than large equipment."[27] Of course, one way to get people was to build a spectacular instrument.

Berkner and AUI were hardly unaware of the appeal that student recruitment and training held for potential patrons of a national facility. Berkner and Tuve both used recruitment to argue for radically divergent views of a national facility. AUI argued that the NRAO wasn't for the leaders in radio astronomy, but for students and new researchers in the field. "To attract young American scientists for training in this field of research and to stimulate new centers of training and research at the universities," said Berkner, "there must be accessible to American science facilities for radio astronomical observation at least comparable to those abroad."[28]

Bart Bok's separate evaluation of AUI's proposal reached the same conclusion: University-based radio astronomy programs were sites for the "training of future PhD's that will be needed by the large facility."[29] Berkner attempted to placate the fears of those at the universities. The NRAO, he said, would "not eliminate the need for smaller installations of the university type, several of which now exists." An early meeting of the Radio Astronomy Committee at AUI's headquarters in the Empire State Building approved the initial plans for the NRAO largely on the basis of the argument that a national facility would provide extra incentives for universities to train radio astronomers. The meeting's consensus saw university-based programs defined "as a means of training," and hence the universities' own radio astronomy programs would "be stimulated by the existence of the large facility" and "would not dry up support for smaller university installations, although in the initial fund raising phases it might be competitive."[30]

All the talk of training, of course, served to bolster funding, but it also exposed the dearth of qualified practitioners ready and able to use a large instrument. In 1955, Otto Struve, later to be named director of the NRAO, asked "Do we imagine any first-class astronomers to use it?" Struve argued that Berkner himself not only should provide leadership during the construction phase of the national facility but also should "promise to personally guide the research."[31] Leo Goldberg of the University of Michigan maintained that "the existence of a national facility will certainly inspire many young men to enter the field." Goldberg anticipated that the national facility, as proposed, would compete with the universities for personnel. But the alternative— limiting the size of the instrument—would defeat the purpose of the institution.[32] Finally, Greenstein suspected that AUI would soon learn "how short we are in personnel, and especially in personnel for the leadership of the group . . . and I doubt whether if it were just built, the suitable people would appear to use it."[33]

Only after concerns about the lack of expertise in radio astronomy had been voiced did the issue of the size of the national facility's centerpiece instrument arise. In July of 1954, AUI called together Lloyd Berkner, Donald Menzel and Bart Bok of Harvard, Armin Deutsch of the National Science Foundation, John Hagan of the Naval Research Laboratory, John Kraus of Ohio State University, Jerome Wiesner of MIT, and H. E. Wells, who represented Merle Tuve and the Carnegie Institution. After general agreement that the AUI proposal took advantage of the knowledge and experience of the antenna projects in Australia and England, "the consensus was that a paraboloid is not a highly speculative device and a high gain, high resolution antenna is certainly a keystone for the proposed facility."[34] Tuve remained opposed to dramatic increases in the antenna's diameter, continuing to argue (and Ira Bowen of Caltech continuing to agree with him) that small stepped increases were the better and safer course. Ira Bowen cited the "near fiasco of the English development" as evidence that "the hazards of progressing by too large jumps without the benefit of experience gained from the intermediate steps."[35] Directly copying the British example was probably unnecessary, Tuve argued; indeed, it would be ill-considered, insofar as the Jodrell Bank instrument was subject to continual structural, feed, and electronic problems. Tuve sought instead to "deflect Berkner" toward considering how the designs of new instruments might be based on a "program of actual radio astronomy observations." A dish of intermediate size might serve radio astronomy better. "If the AUI would request a 140-foot dish," Tuve told Caltech's president, Lee DuBridge, he would be far more sympathetic to the idea of a national facility.[36]

In the course of a long discussion about student recruitment and training, and much debate over size, a generalized critique of "national" science slowly emerged. Berkner himself disliked the label "big" for the proposed National Radio Astronomy Observatory, preferring "national." He had clear, focused ideas about the scope of a national facility, and about its place in the world of science. Playing off a comment made by the Caltech astronomer Rudolph Minkowski about the unity of science, Berkner argued that national facilities did not privilege any one aspect of astronomy but instead allowed "the application of all forms of science . . . in their bearing upon astronomy." Berkner sought to celebrate American science's move toward national facilities by reminding his audience that, although the NRAO was ostensibly concerned with radio astronomy, in fact the "national" label ensured that the enterprise's interests would be kept extremely broad and inclusive. Finally, and most important to Berkner, the label "national" had become, above all, "a means of judging the significance of the kind of research that goes into (or comes out of) the operation."[37]

In contrast, Merle Tuve thought the size of the instrument was almost the only point of the exercise. Tuve was firmly convinced, and Vannevar Bush tended to agree with him, that most of the work in radio astronomy would be "unexpectedly direct and simple." These two towering figures of American Cold War science simply didn't think a huge single dish would provide that opportunity.[38] The historian Allan Needell's work on the confrontation between Tuve and Berkner over the NRAO portrayed Tuve as adhering to a style of "little science"—a style largely idealized in contrast with Berkner's vision of big science under the federal umbrella.[39] The clash between "big science" and "little science" stemmed from larger conceptions of the role of science in the Cold War era. While Tuve largely viewed federal funding for the basic sciences as undermining science itself, Berkner became increasingly convinced that science should serve the state, particularly in times of crisis, and should be willing to be fully supported by government.[40] Deeply impressed by the power and potential of the federal patron to organize large-scale research, Berkner looked to government "to spend literally millions of dollars on a single research tool."[41] Berkner's victory over Tuve in the matter of the NRAO, Needell argues, rippled though the expansion of late-twentieth-century scientific instruments and institutions.[42]

Berkner may have gotten his big dish, but it came with several concessions to the communitarian style of science so passionately defended by Tuve. In fact, the group most responsible for the final shape and style of the American national radio astronomy instrument remained the National Science Foundation Panel on Radio

Astronomy. Acting as a community, the members of the panel were able to get the diameter of the initial instrument reduced to 140 feet—100 feet smaller than Lovell's dish in Manchester and 70 feet smaller than Australia's Parkes radio telescope—and to extract greater support for small university-based radio astronomy programs. Both of these results flew in the face of the fundamental tenets of a national facility, indeed of "big science" as we understand the term. A national instrument for radio astronomy was supposed to be superior to any other instrument in the country and thus to reduce the need for local instrument programs. Local institutions, in turn, could focus on staff development. What actually happened, however, was not that training programs were altered to serve a national instrument. Rather, the national instrument's size and shape were affected by the demands of training future radio astronomers and by the immediate acquisition of expertise.

In April of 1955, Bart Bok, in his capacity as chairman of the AUI Panel, prepared a report titled "Research Objectives for Large Steerable Paraboloid Radio Reflectors." Bok agreed with Tuve and made the choice of instrument dependent on a firm research program, specifically one focusing on the "problems of radio spectral classification," which were "among the most critical of radio astronomy." For John Hagan, another NSF panelist, any investigation to determine "the spiral structure of the gas clouds in the galaxy" required the greater resolution that larger-aperture, dish-type radio telescopes were expected to provide. Though a "special problem" of "extreme importance," galactic structure remained barely touched by the resolution of current apertures, Hagan insisted. Only a "150-foot antenna" able to "increase by a factor of three today's best resolution" could resolve the sky sufficiently.[43] Bok and Hagan defended the choice of the paraboloid as the centerpiece instrument of the national facility by arguing that any objections to paraboloids should "apply equally to many instruments of modified design." There was, however, an important caveat to Bok's continued allegiance to paraboloids. Bok emphasized "the need for precision equipment of large aperture," particularly at 21 centimeters (the wavelength of hydrogen) and at 10 centimeters because solving galactic structure required "precise positioning."[44]

American astronomers supported the proposal for the "intermediate, 150 foot dish" as a "safe step in the realization of the U.S. National Radio Observatory."[45] Though it would be 100 feet smaller in diameter than Manchester's dish, and 60 feet smaller than Australia's, Jesse Greenstein agreed that a 150-foot instrument should present "few problems . . . in construction," being "a conservative step forward over those already built." Greenstein noted that the NRAO dish would still "be the largest single precision paraboloid in this country, and presumably could become an all purpose instrument for use by all cooperating groups."[46] Though Greenstein and Minkowski believed that the expansion of radio astronomy facilities had occurred too rapidly, and largely at the expense of university-based projects, the pair conceded that there were recognizable research objectives for the 150-foot dish. Greenstein sought "excellent detailed information . . . [about] hot gases of ordinary emission nebulae . . . and at high frequencies. [It] is undoubtedly true that a better picture of the total amount of material in such nebulae as the Orion nebula can be better obtained from this device than from any optical technique I know."[47] Even Tuve said he was satisfied that the NRAO would begin securely with an intermediate-size telescope and a formal plan for research.[48] Presumably for the sake of unity, all participants had put the issue of manpower aside for the time being.

Late in July 1956, a day before the choice of the Green Bank site was announced, Jesse Greenstein sent news of the final meeting of the NSF Radio Astronomy Panel to Donald Menzel at Harvard. With Merle Tuve absent, he reported, the mood was of "friendliness and compromise." The "final decision" evidently satisfied all of the participants.[49] The new National Radio Astronomy Observatory would be a steerable parabolic reflector 140 feet in diameter. After that decision, AUI considered parabolic dishes to be the only kind of antennas appropriate for the national facility. The formal declaration noted that large steerable paraboloids "will henceforth be referred to as radio telescopes."[50]

As soon as AUI and the NSF agreed on the size and shape of their new radio telescope, some advisors' thoughts immediately turned to the next dish. Almost immediately, the Harvard physicist Edward Purcell, a Nobel laureate and a co-discoverer of the 21-centimeter line, remarked that the "clamor for increased apertures" was as unlikely to subside in radio astronomy circles as that nuclear physicists would remain "content with any given number of [billions of electron volts]" in new cyclotrons. Purcell, "not convinced" that the "next major investment after the 140 ft. telescope should or will take the shape of a much larger instrument of the same type," speculated that "something much closer to a transit instrument, solar specialists to the contrary notwithstanding, may well yield the biggest scientific returns for a given total investment." Purcell expressed his "disappointment in the AUI study," particularly "its preoccupation with dishes and mounts." Willing to "bet even money" that an improvement on the order of 10 decibels would occur within three years, he argued that AUI should "not assume that dishes and mounts will keep the observatory in the forefront of the science." "The future," he continued, "is not going to be that dull." In other words, rather than seeing the paraboloid as the end point of the development of the observational techniques in radio astronomy, Purcell believed it might be only the beginning. "I'm all for big dishes too," he commented, "but I find it hard to decide whether to allow 1/8" or 1/16" sway in the feed support, when I am not sure we may not want a small cryogenic laboratory mounted out there on the end of it."[51]

The ink was hardly dry on the agreement to build a 140-foot dish before there were calls for a 300-foot dish and even a 500-foot one. Both the 300-foot dish and the 500-foot dish had supporters other than Lloyd Berkner and his colleagues at AUI. John Hagen railed against Merle Tuve's claim that "the proposal for a very large dish . . . is a project of uncertain value." Only a very big dish, Hagan argued, could be suitable for "a long and fruitful program in astronomy."[52] "There can be no doubt that the big antenna will in itself uncover a new class of astronomical problems," Hagan assured the NSF Panel.[53] Greenstein thought it was too early to envision "proper research objectives" for instruments with apertures greater than 500 feet. In marginal annotations to Hagen's letter, he questioned the "newness" of Hagen's claim concerning the possibility to study clouds of neutral hydrogen in emission, but agreed that the "true structure of the hydrogen in the arms of the galaxy" remained a major astronomical question.[54]

Hagan's support for a 500-foot parabolic dish reopened the debate between advocates of parabolas and advocates of interferometers. Hagen emphasized the benefits of the large parabolic reflector, particularly its "high resolution" and "greater flux gathering power," and criticized interferometers at centimeter wavelengths:

. . . with the many fringes of the interferometer the "lumpiness" or fine structure of the galactic background will appear as background "noise" effectively masking weak sources. With both arrays and interferometers of wide spacing the bandwidth that can be used is seriously restricted because of lack of signal power. This of course means that sensitivity is automatically reduced and long integration is required in an attempt to regain it. Any cost comparisons of arrays of antennas as compared with single large reflectors must take into consideration not only these limiting factors but also the fact that the array (such as a Mills Cross) is a single frequency device and if one wants information over the spectrum he must make many antenna each of limited usefulness. Such antennas have a place in the scheme of things but they do not lend themselves to making the detailed studies possible with the large reflector.[55]

Hagan's critique of interferometers at centimeter wavelengths serves to remind us that with the identification and importance of hydrogen's 21-centimeter line the interferometer's former greatest weakness—frequency specificity—became its greatest strength. As the panel considered the form of future radio telescopes, the interferometer once more appeared to rival the parabolic dish. The University of Michigan astronomer

Leo Goldberg argued that only a large parabola could rival the resolution of an interferometer. A very big dish would also address the major shortcoming of inter-ferometers: their tendency "to become confused by multiple sources." "Really effective observations of radio noise from small areas on the sun can only be carried out with large dishes," Goldberg asserted, "and in this respect even a 500 footer is none too large."[56] As Edwin Purcell astutely recognized, in the Cold War world almost nothing would satisfy scientists' desire for ever-larger instruments.

By 1955 or 1956, Greenstein must have recognized that the sheer cost of proposed gigantic dishes was probably out of reach of even the federal patron. Extrapolating upward from the cost estimates for the 140-foot NRAO dish and Jodrell Bank's 250-foot Mark I, Greenstein concluded that a 600-foot solid dish would cost about $75 million. He concluded that it was "unwise to budget now."[57] As we now know, only the military accepted plans for a 600-foot dish, although not for radio astronomy but for the purpose of listening in on Soviet radio signals reflected off of the moon at a separate site called Sugar Grove. Whereas funding for the Sugar Grove project came from the Office of Naval Research and the Naval Research Laboratory, funding for the Green Bank project came entirely from the National Science Foundation. The former was complex-style science, militarily funded and militarily applied. The latter, community-style science, focused on pure science and preoccupied with the training of future radio astronomers. Optimistically, AUI officials believed that the two projects should be "kept in touch."[58]

Successfully reducing the size of the initial radio telescope for the NRAO was only the first victory for community-oriented astronomers. The very existence of a "national" facility continuously threatened future NSF allocations for university-based science. Rudolph Minkowski and Jesse Greenstein agreed that "training of students and research personnel, current support of ongoing research projects, and new capital expenditures at universities active in radio astronomy" should be the priorities for support garnered from government sources.[59] As Bart Bok supported AUI's continued efforts to plan, build, and staff the National Radio Observatory, Greenstein wrote an emphatic "NO" next to each of Bok's suggestions. In response to Bok, Greenstein and Minkowski cited the example of Brookhaven National Lab to portray the destruction of university-based radio astronomy if anything more than a skeleton staff were to be employed permanently at Green Bank:

. . . if one studies the ratio of man years of research carried out by visitors to Brookhaven as compared to that by the permanent staff it seems to be near 1/10. The active radio astronomers in the U.S. are so few that such a ratio would destroy all going institutions in the field. There would therefore be a serious defect in future training, and the university contribution would become effectively zero.[60]

In other words, following the model of Brookhaven would fundamentally destroy radio astronomy, since new students could only be trained at universities and the national facility would eventually eliminate the universities' radio astronomy programs. Additionally, it was the large number of physicists, particularly particle physicists, that had created the necessity for a national laboratory at Brookhaven. There was no comparable "backlog of personnel available" in radio astronomy; indeed, "training new men" was difficult.[61] From Greenstein's perspective, the demands for better and substantially larger instrumentation in radio astronomy paled before the demands of student training and staff expansion at universities. Greenstein's challenge was so strong that Bart Bok revised his plan; he now advocated "partial support to the centers . . . like Ohio State, Harvard, and Cornell." The two priorities for radio astronomy in the United States remained, Bok said, a "steady research productivity . . . at Universities" and "a steady flow of PhD's with the required background of cross-field training."[62]

And so the AUI Planning Document changed once more. The new proposal, submitted to the NSF in late 1956, outlined the general argument for a National Radio Astronomy Facility and its instrument and staff requirements, but now specified the facility's relationship to university-based researchers. Though the document acknowledged only four active research programs in the United States, encompassing fewer than thirty participants, Berkner and AUI continued to argue that the "growing deficiencies of research facilities" remained the major impediment to American "leadership" in the "rapidly developing field."[63] The new national facility would now fulfill four specific objectives. First, it would allow the building of unusually large instruments to address established research questions. Second, it would "integrate optical and radio studies more effectively," since each specialization would have equal access to the radio astronomy instruments. Third, it would "encourage universities and other research institutions to plan radio astronomy projects of their own" without the need for large expenditures on equipment. Finally, "a National Radio Astronomy Observatory will be invaluable in the training of students." AUI and the NSF thus emphasized the need for recruitment and training in astronomy. Moreover, they accepted the changing disciplinary identity of a field that required, in equal measure, "thorough knowledge of both optical astronomy and of the special techniques and problems of radio astronomy."[64]

It was this measure and meaning of a "national" radio telescope that received funding. For fiscal year 1957, the National Science Foundation requested and received

$4.5 million for an "Inter-university Radio Astronomy Facility," maintaining that the United States ("birthplace of radio astronomy") was "lagging far behind other nations" and citing the giant radio telescopes under construction in Australia and England.[65]

The National Radio Astronomy Observatory was, in the end, corralled into being a more community-style science than was first envisioned. Indeed, by 1957 the NRAO more closely resembled the university-based radio astronomy programs in the United States. This isn't surprising, since many of the participants were involved both in the NRAO and in university-based programs. Crucial to the NRAO's establishment and support from the astronomical community was the acceptance on the part of Berkner and AUI of two fundamental truths about American radio astronomy: that American radio astronomers were more closely tied to astronomy than Australian or British radio astronomers were and that the astronomy community sought to expand its appeal across multiple disciplines in order to attract students.

For the East Coast institutions, there were already signs of a payoff by 1956. Bart Bok reported "the terrific boon to the development of student interest in the Eastern United States in radio astronomy that has already come as a result of the preliminary announcements regarding the Green Bank operation." He claimed to have "twelve young people" actively working at Harvard's Agassiz Station and another dozen expressing interest in radio work in their applications to graduate school.[66] Bok's colleague Donald Menzel anxiously awaited approval of AUI's plan's approval, official approval of the site, and construction. Radio astronomy was genuinely "going to need the facility," Menzel said, because Harvard's students were "coming out." Harvard's graduates, at least, would "be able to staff that observatory very well by the time it gets built."[67]

Tuve advocated delay, but Greenstein, though cautious, pushed "to get our feet wet." Only if construction were begun would problems be encountered and solved. As to the manpower problem, Greenstein cited the example of the British and Australian radio astronomers, whose success was "based on the work of young men." Greenstein saw "young men coming along" but acknowledged that a majority of the NSF Panel still believed that the United States lacked "the ideal leader" for the national facility—and he claimed "not [to] be embarrassed at importing one."[68] The leader's of the United States' "national" facility, no longer possessing the largest dish and not sure exactly whom it would employ, now faced the possibility that there simply was no American who was qualified to lead the project.

Menzel wrote to Ira Bowen asking him to serve on the committee charged with selecting a director for the National Radio Astronomy Observatory. "The primary restriction," he noted, "is that the nominee should be an American citizen." The committee was to have discretion as to "whether this director should be primarily an astronomer or primarily a radio engineer." Menzel's position was that "direction of the scientific program, in the long run, is more important than ability to design electronics."[69] Bowen maintained that the NRAO's director should have the multiple layers of divergent interest that embodied the radio astronomy community, including "knowledge of and research ability in general astronomy," "familiarity with electronic techniques," "experience in radio astronomy," and "executive ability." What remained most important to Bowen, however, was that "the most fundamental contributions to astronomy by radio techniques" had come from observatories at which an astronomer rather than a radio engineer had run the program. Bowen cited Harvard and Leiden as the best examples. He believed that this pattern would continue, and that it favored an emphasis on research ability in general astronomy. That left Bowen and Menzel with only one clear choice for a director: Bart Bok, whose unhappiness with Harvard had become well known and who seemed increasing likely to leave for Australia.[70] John Hagan at the NRL agreed that the choice of Bok would make the NRAO "a true center of research."[71]

When Bok decided to go to Australia's Mount Stromlo Observatory, the requirement that the director be an American citizen became a contentious issue. "It would look mighty queer," Merle Tuve commented, "if among American astronomers, we could not find someone with proper qualifications," since "the justification for support by the National Science Foundation was the need of <u>American</u> astronomers for the radio facility."[72] Menzel suggested three Australians, (John Paul Wild, Edward Bowen, and Joseph Pawsey) and one Englishman (Robert Hanbury Brown) if the citizenship requirement could be relaxed. Otto Struve suggested Pawsey; Jerome Wiesner suggested Edward Bowen, John Bolton, or Bernard Lovell.[73]

Leo Goldberg was offered the directorship and declined it in late 1956. Berkner and AUI had appealed, evidently, to Goldberg's "sense of duty," which Goldberg declared was not enough to persuade him to leave the University of Michigan. The appointment to a national facility, Goldberg retorted, would limit his scientific freedom "to pursue any line of activity that [he considered] personally rewarding." The directorship would force him to abandon optical solar physics for radio astronomy. The university, rather than the national facility, was the best venue for scientific freedom, he argued. Green Bank looked to be in "relative isolation" from the universities and was just not "the real thing."[74] (As it turned out, Goldberg's commitment to the University of Michigan was tested in 1958 by an offer to return to Harvard as Bart Bok's replacement, which he accepted.)

In 1959, Otto Struve accepted the directorship of the NRAO, hoping to make a success of a move from stellar spectroscopy to radio astronomy. By the middle of 1961 he had already tendered his resignation twice, the second time on medical grounds. He would later claim that he stayed on after Berkner's resignation from AUI and desperately attempted to get completion of the 140-foot telescope back on a reasonable schedule. In fact, he came to the realization that the 140-foot dish was increasingly pointless, especially after Jodrell Bank and Caltech beat him to his goal of finding long-wave radiation from optically observable stars. Moreover, he claimed that his other important anticipated project, obtaining closer correlations of optical and radio stars, "cannot be done with the present equipment."[75]

In mid 1961, whispers of criticism of the 140-foot dish began circulating around AUI. The radio astronomer Frank Drake, in the same year in which he developed his famous equation for estimating the probability of interstellar contact, expressed a belief that the 140-foot dish would be "useful only for measurements of radio fluxes and positions for radio sources at wave lengths shorter than about 10 cm." At such wavelengths, "emission from clouds in the earth's atmosphere causes serious distortions." The most effective solution was to shift the site of the telescope out to Kitt Peak or some other desert site with less atmospheric moisture.[76] Even Lloyd Berkner had recognized a year earlier that the 140-foot telescope's "uniqueness" had diminished. Berkner claimed, however, in words echoing Alvin Weinberg and "big science," that the "140-foot telescope has become a symbol in the eyes of the scientific community generally, including members of the NSF Board and to the Congress and the public which is, of course, the ultimate source of support for such an institution as the NRAO."[77] Even the smaller dish, Berkner believed, still evinced the American effort in leading the field of radio astronomy.

Struve, as an active astronomical researcher, didn't adjust well to the job of directing a national laboratory. Near the end of 1961, he requested an extended leave of absence, perhaps several months, "to give me an opportunity to engage more actively in research. . . . I feel that I must try to catch up with recent developments in astrophysics and I am unable to do so while I am compelled to spend nearly all of my time in non-scientific meetings." Struve blamed the bureaucracy for his fatigue and related "health problems."[78] Leo Goldberg's assessment of the NRAO became bleaker. A distinct "absence of leadership" had emerged by the end of 1961, he said. Construction of the 140-foot telescope continued to drag on. The director was "simply marking time," chiefly concerned with the "protection of his personal reputation in absolving himself of any blame for the failure of the 140' telescope." Looming over it all were the emerging plans for a 300-foot telescope, estimated to cost $13 million. The scientific staff, Goldberg told AUI's new director, Isidore Isaac Rabi, had "serious doubts" that the cost could be justified on "scientific grounds."[79]

Late in 1961, it was announced that Struve would be succeeded by the Australian radio physicist Joseph Pawsey. Any respite from criticism was short-lived; only a few weeks after Pawsey arrived in the United States, he was dead from a cerebral hemorrhage. The new choice of a director—David Heeschen—was the final act in the saga of the NRAO. Back in 1954, Heeschen had been the one and only American-trained radio astronomer that Bok, Tuve, Greenstein, or even Berkner could identify as presumably requiring a massive federally funded radio telescope. And now, at last, he had come.

The Dish, a film featuring the Parkes radio telescope and telling the story of how radio astronomers captured the television images of the moon landing in July of 1969, was a comic hit in Australia in 2000. At first the Parkes telescope was scheduled to be only a backup receiver for the moon mission, but Neil Armstrong's decision to forgo a scheduled sleep break put the Parkes dish in a favorable position. At the moment Armstrong set foot on the moon, the moon was visible from Australia but not from Houston. In the film, the prime minister, speaking before Parliament, lauds the achievement of the Parkes dish as perhaps the proudest moment in "Australia's scientific history." His private reaction to the news that Australia "got the moonwalk" is an all-too-Australian mixture of surprise, concern, and pride: "Shit!" Indeed, much of the film's humor comes from the awkward participation of ordinary citizens of Parkes in an international spectacle. The otherwise lighthearted film concludes dramatically. Just as Armstrong begins to exit the landing module, freakish gusts in Australia place wind loads on the telescope. While people all around the world stares at "snowy" television images, tension builds as the small group of heroes fight to capture the images better.

Early in the film, Prime Minister John Gorton asks his aide what the dish is "doing in the middle of a sheep paddock."[1] In the lore of the radio astronomers, spending years in one sheep paddock or another resonated with the struggle to create a new intellectual and social community of science. For the Australian radio astronomer Edward "Taffy" Bowen, sheep represented the narrow nationalistic and practical goals of many scientists and politicians. In another way, sheep in a paddock became a metaphor for the cluelessness of most people about the work of radio astronomers. After being surrounded by sheep for many years at Jodrell Bank, Robert Hanbury Brown moved to the Australian town of Narrabri to build his stellar interferometer. "If you could stand the heat in summer," he recalled, "played reasonably good tennis and liked to talk about sheep, wheat and the rainfall," Narrabri was a good place to live.[2] Scientists exist in intellectual and social communities, often rather at odds with the provincial and bewildered yet proud neighbors who house and support them.

The Dish evokes many of the themes of this book. During the Cold War, the emergence of radio astronomy revolved around the interplay of local, national, and international interests, both of people and of science. That the moon wasn't physically visible from Houston at a critical moment in July of 1969 nicely underscores the irrelevance of national boundaries to the scientific community. Yet the fact that the vehicle that had landed on the moon was American highlights the national immediacy of scientific endeavors. The world watched the moon landing on television; everyone with a television saw the same images and celebrated Armstrong's transcendent claim that the moon mission was "for mankind." Radio astronomers played a small part in that grand spectacle, having built a worldwide network of radio telescopes. That such a network existed by 1969 is a testament to the radio astronomers' international and interdisciplinary scientific community.

This book has inverted the familiar story of Cold War science, according to which the new and intimate relationship between science and the state seemingly dominated the process of knowledge creation. The radio astronomers refused to be reduced to a handmaiden or a servant of national policy; they already saw the era's most glamorous science, nuclear physics, in that role. Though individual radio astronomy groups and facilities certainly germinated within local disciplinary and national scientific cultures, the radio astronomers gained resources by arguing that modern nations were members of the international community and, likewise, that scientists were members of larger interdisciplinary communities. Those twin interrelated rhetorical and real communities forged the measure and the meaning of radio astronomy.

The galactic astronomer Bart Bok concluded that radio "change[d] the face of astronomy."[3] After 1945, community transformed radio physicists into radio astronomers, laboratories into observatories, antennas into telescopes, and noise into vision. Radio and electronics were among the new technological opportunities that astronomers sought immediately after World War II, along with space-based astronomy, photoelectric detectors, and consortia capable of building larger observatories.[4] Pursuing new instruments, new disciples, and other wavelengths, the radio astronomers helped create modern all-wave astronomy. All-wave astronomy uncovered a new cosmic horizon and the 21-centimeter line. Later it would add pulsars, quasars, cosmic background noise, and evidence of the big bang to astronomy's stock of achievements.

Astronomy changed when the 200-foot radio telescope at Jodrell Bank became analogous to the 200-inch Palomar Mountain optical telescope in California. Observatories employed radio physicists, while radio physics laboratories sent graduate students to learn astronomy. Both optical astronomers and radio physicists adjusted their expectations of astronomical standards and even their conceptions of normal astronomical objects to meet the demands of the new science. A radio telescope became a normalized part of astronomical technology, while the optical telescope was reduced in stature to merely one among many components of astronomical practice. And all members of the new community became, in effect, graduate students. The radio astronomers' intellectual world shaped their social world: an interdisciplinary, all-wave astronomy couldn't exist without a new international astronomy community.

That science could be used to break through social and intellectual boundaries of all kinds might be the most enduring legacy of the radio astronomers. The story of the radio astronomers adds an important new dimension to the history of science in precisely this way. From the Renaissance through the Enlightenment to the nineteenth century, as much recent work has amply demonstrated, esoteric ideas were always firmly grounded in practical problems—fortification, cable telegraphy, clocks, and so on.[5] In more recent science, seemingly arcane fields of knowledge such as radio astronomy confronted pressing social problems and offered uncomfortable solutions. By the 1970s, the conference that first used the term "all-wave astronomy" looked to conceptualize the techniques aliens might use to communicate with humanity. What was this really about? In fact, for the assembled astronomical community, all-wave astronomy was a brave attempt to solve one inscrutable social and political problem right here on Earth. If astronomers could learn to communicate with aliens, capitalists could learn to communicate with communists.[6]

The radio astronomers' emphasis on community, I have argued, can be understood only in contrast to the abundant discipline of the physicists. Since 1945, the physicists (and their historians) have been "thrilled" by "all the bright new chromium plated gadgets" on offer during the Cold War. The history of science and technology in the Cold War era has long revolved around the increasing abundance of gadgets, budgets, and students. Occasional voices noted how "chromium plated gadgets" had lured physicists into industry or governmental laboratories, where, as in wartime, they did little in the way of actual physics. Among those voices was Caltech's president, the physicist Lee DuBridge, who noted how "a portion of the prestige which physicists have acquired is based on a misunderstanding of their normal function." The physicist "is not an inventor of gadgets or weapons but is one who seeks knowledge and understanding."[7] Nonetheless, many physicists took the lure of the military-industrial complex and were caught. Only in later decades would they realize that, though they gained the wealth of Solomon from the military-industrial complex, it didn't make the chrome any less a veneer.

Glamorous stories of nuclear physics have obscured the fact that there were alternative communities of scientists and alternative ways of doing science during the Cold War. Other scientific communities sought a broader counter-culture of scientists and engineers who envisioned a different path for knowledge and for humanity.[8] The radio-physicists-*cum*-radio-astronomers continued a model of interdisciplinary and international community of science that had been pioneered during World War II. One larger argument of this book has been that the radio astronomers inverted the standard technocratic superpower metanarrative of the Cold War state, with its secrecy, its parochial material immediacy, its rampant nationalism, and most of all its delusional bluster.[9] I have argued that community-style science triumphed over complex-style science in the case of radio astronomy—most evidently in the case of the National Radio Astronomy Observatory. Even when the model of complex-style science was adopted by the radio astronomers, it only served to push a community-style approach. The artificial requirement that an American be selected to run the facility was brushed aside in favor of international leadership and cooperation.

The work and the legacy of one of the most famous disciples of the era, Carl Sagan, illustrate both that community styles of science emerged during the Cold War and that they remain very much a part of modern all-wave astronomy. Sagan sought to fulfill the ideal of a scientific community uncorrupted by military applications and nationalist concerns.[10] Like many scientists, including the radio astronomer Martin Ryle and the biologist Linus Pauling, Sagan overtly criticized most sciences' relationships to the military-industrial-academic complex in the Cold War world.[11]

If Arthur C. Clarke began the creation of a new astronomy in 1945 by encouraging old optical astronomers to pursue the new technologies that had emerged from World War II, Sagan's goal was nothing short of an expansive, inclusive vision of an all-wave astronomical community at work for humanity. As fellow popular science writers, fellow novelists, and at least on one occasion fellow panel members, Clarke and Sagan overlapped many times in the decades this book has traced. Contentiously for a professional scientist, Sagan spent the last two decades of his career engaged in clear popular exposition of astronomy and science. He produced numerous books, lectures, and a visionary television series, *Cosmos*, that conveyed the awe and majesty of the unfolding astronomical universe, helped along by the space probe Voyager II's color images of Jupiter and Saturn and his deeply resonant voice.

The radio astronomers established an exemplar for Sagan's understanding of what science was. He understood that radio, microwave, millimeter, and infrared wavelengths were as important as optical wavelengths to the project of revealing the mysteries of space. As a faculty member at the Smithsonian Astrophysical Observatory at Harvard in the generation after the widespread changes to astronomy this book has charted, Sagan published prodigiously. Many of his publications were on controversial topics, such as "higher organisms on Mars" and "exotic biochemistries in exobiology."[12] As early as the mid 1960s, with the radio astronomers a coherent intellectual and social community, Sagan could work as easily in radio astronomy as in exo-biology; he pursued cosmology and planetary astronomy. He helped organize the conference that first used the phrase "all-wave astronomy," which succinctly describes the complete transformation of the measure and meaning of astronomy as "astronomer" and "tele-scope" became categories inclusive of radio physicists and optical astronomers, of electronics and of lenses, and of scientists from all nations.

In concluding this book, it seems appropriate to continue the theme of cross-boundary cooperation and communication by looking at Sagan's first novel, *Contact*, which insightfully explicates the competing social arrangements—community versus complex—of Cold War science. The plot of *Contact* revolves around the reception and deciphering of a Message of unknown extraterrestrial origin. A Machine is con-structed on the basis of plans found in the Message. Sagan explores how the discovery of such a Message from the stars might affect human institutions in the mid 1980s. He envisions the Soviet Union and the United States reducing their nuclear arsenals, millennial cults abounding, widespread feelings of expectation and optimism, new technologies and industries, and a far greater role for the United Nations.[13] Social progress, he suggests, would follow from inclusive, increasingly international and trans-disciplinary science. Like the astronomers immediately after World War II, Sagan didn't reject technology. Instead, he believed that the most significant changes would be found in new social arrangements spurred by new communities' seeking new uses for new technologies. In that way, Sagan evoked the lofty hopes of the disciples of radio astronomy at least as far back as the 1950s.

In *Contact* Sagan exposed a quandary of Cold War science: If you have limited technical expertise, and commit most of it and your resources to something mundane such as national security, you cannot hope to be prepared and trained to receive a message that will change humanity. "There are at most a few hundred really capable radio astronomers in the world," a Russian character says to his American counterpart.

"That is a very small number when the stakes are so high. The industrialized countries must start producing many more radio astronomers and radio engineers with first-rate training."[14] With its emphasis on recruitment and training, the identity of the radio astronomer encapsulates the history of the first two decades of the creation of the international radio astronomy community. Likewise, Sagan's emphasis on science's internationalism (the Message is decoded and Machines are built by worldwide consortia) replays a feature of the radio astronomers' community. To drill that message home, Sagan concludes the novel with five human travelers crossing the galaxy in an inconceivably short time and returning home with nothing of recognizable worth to the military-industrial complex, only a message of interstellar peace.

Near the end of *Contact*, Dr. Eleanor Arroway,[15] Sagan's heroine, asks an extraterrestrial about the vast amount of matter being consumed by two black holes at the center of the galaxy. In a clear nod to the history of radio astronomy, the extraterrestrial explains that the object is Cygnus A, the very first galactic radio source identified by John Bolton back in 1948. In Sagan's imagination, Cygnus A is not really a natural object but evidently is part of an experimental galactic cultivation program run by a cooperative consortium of species and galaxies.[16] Sagan's vision of the extraterrestrials' ambitions in making contact with an Earth-level culture reveals, of course, Sagan's own expectations about his own culture. Secure in the traditions of science fiction, his purpose is not to reveal the mysteries of E.T., but to comment and reflect on the place and purpose of science in his world. Science, Sagan suggests, *can* be used cooperatively and internationally to advance humanity. Thus Sagan emphasizes "community" and pictures a transformed post-Message world in which "distinctions that had earlier seemed transfixing—racial, religious, national, ethnic, linguistic, economic, and cultural—began to seem a little less pressing."[17] In other words, community-style science must struggle to elevate inclusive, open, international, and interdisciplinary knowledge in the face of complex-style science (defined as closed, secretive, nationalistic, and subservient to the narrow ambitions of industry and the military). That is urgent, Sagan repeatedly and sadly notes, because most science is complex-style science, used merely to help us kill ourselves.

Sagan's novel illuminates how some Cold War–era scientists interpreted the broader struggle between community-style science and complex-style science. The novel's thematic concerns demonstrate that the competing social modes of community and complex pervaded the changing conduct and character of science during the Cold War. The case of the radio astronomers was not an isolated one. Consider the budget priorities of the National Science Foundation for 1957, the year of Sputnik, a noted moment in the history of complex-style science. In that year, the NSF requested more than \$15 million from Congress to support "productive basic research." The NSF targeted four areas for equipment and facility development: astronomy, nuclear research, electronic computation, and biology. Radio astronomy took the largest portion of astronomy's allocation, nearly 30 percent, the majority being spent on the new National Radio Astronomy Observatory. Meanwhile, "University Nuclear Reactors," "Computing Facilities," and "Biological Field Stations" received about \$1.5 million apiece to be spent over the period 1957–1960. Receiving \$570,000, the final category of Biological Field Stations contained a line item for "Controlled Environment Laboratories," literally computer-controlled greenhouses but better known by the alluring name of "Phytotrons."[18]

What is fascinating about the funding priorities of the National Science Foundation in 1957 is the absence of the usual Cold War scientific objects such as space rockets and cyclotrons in favor of radio telescopes and phytotrons. Those priorities suggest that rockets, cyclotrons, and atomic bombs have served only as a first approximation to the history of science since 1945. In short, around 1957–58 (significantly, during the International Geophysical Year, yet another international and interdisciplinary scientific effort) more community-style sciences gained substantial patronage even as the Cold War grew hotter. The broader argument suggests that the Cold War era understood normal science as "productive basic research." Fully understanding the practice of normal science in the Cold War era requires parallel narratives for radio astronomy, university nuclear reactors, computing centers, and controlled environments.[19] Histories of these four sciences, contemporaneously and consciously lumped together by the American NSF, the Australian CSIRO, and the British DSIR, focused on a range of new technologies and social organizations aimed at solving the broadest of social problems. University nuclear reactors were aimed at training scientists, engineers, and technicians to build and run nuclear power plants for industrial and social development all around the world. Phytotrons promised universal plant species adaptable to any environment. Radio astronomy heralded open communication among all peoples.

There is a history to be told of Western nations in the Cold War supporting international and interdisciplinary sciences. Whether we look at American, British, French, or Australian funding priorities for science and technology around 1957, the narratives of radio astronomy, computing facilities, university nuclear reactors, and controlled environment laboratories appear strikingly similar. Australia had begun to build the Parkes radio telescope and had seen the Lucas Heights nuclear reactor become operational, and the Australian government had just approved millions in funding for a phytotron. In France, an early adventure in radio astronomy had led to a cooperative venture headed by the Netherlands, a phytotron was built outside Paris, and a central-ized computer project was begun. In the United Kingdom, the University of Man-chester had built and run one of the earliest electronic computers, the Universities of Manchester and Cambridge had erected large radio telescopes, and Imperial College London had bid successfully to administer Britain's first "teaching reactor," intended to train the teams of nuclear engineers that a large-scale nuclear power industry would require.[20]

And so this book concludes with a deeper analogy. During the Cold War, secrecy, divisions, and nationalism became paramount concerns of the military-industrial complex, which in turn gained vast patronage and power from the culture of fear. At the same time, that same context fostered community-style science, in which inter-national and interdisciplinary practitioners cooperated openly, confounding all the supposed boundary lines of the era. Radio astronomy as a field of knowledge chal-lenged the disciplinary fragmentation of science. The radio astronomers shook the social boundaries of scientists, consciously making a new international community. Likewise, the radio telescope itself defied the accepted conception of an astronomical instrument. The story of the radio astronomers opens a window onto an alternative conception of social and knowledge communities during a divisive and suspicious time. Uncommonly parallel cases serve to demonstrate that the radio astronomers were not isolated or unique in their boundary crossing, but instead evoke a broader history of competing models of scientific organization. Moreover, recovering com-munity-style science can work to erode petty divisions of nationhood, religion, or language, can aid communication with others, and can help educate new disciples. Together, those stories from that era serve to guide people toward learning how to live in a global community.

ABBREVIATIONS

AA	Australian Archives
ASRLO	Australian Scientific Research Liaison Office
AUI	Associated Universities, Inc.
BAAS	British Association for the Advancement of Science
CIT	California Institute of Technology (Caltech)
CSIR	Council for Scientific and Industrial Research (till 1949) (Australia)
CSIRO	Commonwealth Scientific and Industrial Research Organisation (after 1949) (Australia)
DSIR	Department of Scientific and Industrial Research (Britain or New Zealand)
HCA	Harvard College Archives
HCO	Harvard College Observatory
HCOC	Harvard College Observatory Council
IAU	International Astronomical Union
NRAO	National Radio Astronomy Observatory (USA)
NRL	Naval Research Laboratory (USA)
NSF	National Science Foundation (USA)
ONR	Office of Naval Research
RAC	Rockefeller Archives Center, Sleepy Hollow, New York
RAS	Royal Astronomical Society (Britain)
TRE	Telecommunications Research Establishment (Britain)
URSI	Union Radio-Scientifique Internationale (International Scientific Radio Union)

NOTES

INTRODUCTION

1. *The Englishman Who Went Up a Hill but Came Down a Mountain* (Miramax, 1995). For a great example of art as life, compare Anson's superior, Captain George Garrad, to Colonel Sir Charles Close as depicted in *The Early Years of the Ordnance Survey* (David & Charles, 1969). See p. 55 of the latter on the measurements of "hills."

2. "Radio Astronomy," lecture to a Communications summer course, August 2, 1956, Jan Oort Papers, University of Leiden library, box 68, folder c, p. 3. The Dutch astronomers readily adopted the rhetoric of revolution to describe the new discoveries of radio. For example, see Per Olof Lindblad, "Early Galactic Structure," in *Oort and the Universe*, ed. H. van Woerden et al. (Reidel, 1980), 64. The topic of resistance or embrace of new technology for the pursuit of science was also apparent when Robert Hooke of the early Royal Society became frustrated at Johannes Hevelius' claims about astronomy without the aid of newer instruments. Hooke charged that with "telescopic sights" Hevelius "would have afforded the World Observations . . . ten times more exact." See Lisa Jardine, *The Curious Life of Robert Hooke* (Harper, 2003), 43–44. Editor's note: In these notes, only main titles are given. Subtitles are given in the bibliography.

3. Robert W. Smith, "Engines of Discovery," *Journal for the History of Astronomy* 28 (1997): 49–77; Richard F. Hirsh, *Glimpsing an Invisible Universe* (Cambridge University Press, 1983); Andrew J. Butrica, *To See the Unseen* (NASA, 1996); David DeVorkin, *Science with a Vengeance* (Springer-Verlag, 1992).

4. Robert W. Smith, *The Space Telescope* (Cambridge University Press, 1993); David DeVorkin, "Electronics in Astronomy," *Proceedings of the IEEE* 73 (1985), no. 3: 1205–1220; W. Patrick McCray, *Giant Telescopes* (Harvard University Press, 2006).

5. David Edge and Michael Mulkay, *Astronomy Transformed* (Wiley, 1976); Jon Agar, *Science and Spectacle* (Harwood, 1998); Peter Robertson, *Beyond Southern Skies* (Cambridge University Press, 1992); J. S. Hey, *The Evolution of Radio Astronomy* (Science History Publications, 1973); R. Haynes, "From Swords to Ploughshares," in *Explorers of the Southern Sky*, ed. R. Haynes et al. (Cambridge University Press, 1996); W. T. Sullivan III, "Early Years of Australian Radio

Astronomy," in *Australian Science in the Making*, ed. R. Home (Cambridge University Press, 1988). Lovell is generally known as Bernard, though his full name is Alfred Charles Bernard.

6. Robert Hanbury Brown, *Boffin* (Adam Hilger, 1991); Bernard Lovell, *Echoes of War* (Adam Hilger, 1991); Edward G. Bowen, *Radar Days* (Institute of Physics Publishing, 1998).

7. Woodruff T. Sullivan III, *Cosmic Noise* (Cambridge University Press, 2009); John Lankford, *American Astronomy* (University of Chicago Press, 1997).

8. Sullivan, *Cosmic Noise*, 47, 29–53, 54–78; Woodruff T. Sullivan, "Karl Jansky and the Discovery of Extra-Terrestrial Radio Waves," in *The Early Years of Radio Astronomy*, ed. W. Sullivan III (Cambridge University Press, 1984).

9. Sullivan, *Cosmic Noise*, 394. Sullivan's excellent use of extensive oral history sources, being deeply indebted to traditions of using historical evidence, stems from his longstanding connection with historians of science. For a more naive account using oral history, see pp. 261–263 of W. Miller Goss and R. X. McGee, *Under the Radar* (Springer, 2009). In their biography of the pioneering radio astronomer Ruby Payne-Scott, Goss and McGee rely on distant charitable memories of the victors in the fight between the two halves of the Australian radio astronomy group to provide a rationale of their work giving only an impression of the commentators themselves, rather than their subject.

10. See *Transactions of the IAU* 8 (Cambridge University Press, 1954), 38 (quoted in Edge and Mulkay, *Astronomy Transformed*, 63).

11. Greater interdisciplinary work in Cold War astronomy has been noted previously, with the important caveat that it is usually dated to the post-Sputnik period. Smith ("Engines of Discovery," 55) discusses this trend for planetary sciences.

12. Thomas S. Kuhn, *The Structure of Scientific Revolutions* (University of Chicago Press, 1996), 8.

13. Arthur C. Clarke, "The Astronomer's New Weapons," *Journal of the British Astronomical Association* 55 (1945): 143–147, quoted in David DeVorkin, "Electronics in Astronomy," *Proceedings of the IEEE* 73 (1985), no. 3: 1205–1220.

14. Eileen Reeves, *Painting the Heavens* (Princeton University Press, 1997), 3–4.

15. Sharon Traweek, *Beamtimes and Lifetimes* (Harvard University Press, 1988), x. See also Albert van Helden and Thomas Hankins, "Introduction: Instruments in the History of Science," *Osiris*, second series, 9 (1994); Nicolas Rasmussen, *Picture Control* (Stanford University Press, 1997). In the creation of new knowledge, I especially value Robert Kohler's work connecting a new type of model organism with a peculiar social community of *Drosophila* geneticists. See Kohler, *Lords of the Fly* (University of Chicago Press, 1994).

16. Noel Perrin, *Giving Up the Gun* (Godine, 2004); Ruth Swartz Cowan, *More Work for Mother* (Basic Books, 1983), 128–150; Terry Reynolds and Theodore Bernstein, "Edison and 'The Chair,'" *IEEE Technology and Society Magazine*, March 1989: 19–28; Rudi Volti, "Why Internal Combustion?" *American Heritage of Invention and Technology*, fall 1990: 42–48.

17. Edward M. Purcell to Merle Tuve, June 27, 1956, Leo Goldberg Papers, Harvard College Archives, box "L. G. Corres. with NRAO," file "Steering Committee—AUI Radioastronomy Facility."

18. In raw terms, up to 1949 some 64 schools produced 1,765 PhDs in physics, and from 1949 to 1958, 81 schools produced 3,375 PhDs (American Council on Education, *American Universities and Colleges*, sixth edition, ed. M. Wilson (American Council on Education, 1952), 54–57; American Council on Education, *American Universities and Colleges*, eighth edition, ed. M. Wilson (American Council on Education, 1960), table 6; Walter McDougall, . . . *The Heavens and the Earth* (Basic Books, 1985), 160–62; JoAnne Brown, "'A Is for Atom, B Is for Bomb'" *Journal of American History* 75 (1988): 68–90.

19. Paul Forman, "Behind Quantum Electronics," *Historical Studies in the Physical and Biological Sciences* 18 (1987): 149–229, 204. Engineering ballooned even more, becoming the largest single occupation for white-collar males. See Matthew H. Wisnioski, *Engineers for Change* (MIT Press, 2012).

20. "A Few Highlights in Teaching and Research for the Division of Physics, Mathematics, and Astronomy" and "Appendix—'Staff Members and Students in Physics, Mathematics, and Astronomy in residence during the year 1956–57," Lee DuBridge Papers, California Institute of Technology Archives, Folder 30.12.

21. In the related cases of radar astronomy and planetary astronomy, it was only post-Sputnik that the military-industrial complex rapidly, and tragically, enticed many new recruits into large, politically driven projects. See Andrew J. Butrica, *To See the Unseen* (NASA, 1996), viii.

22. Numbers complied from data in American Council on Education, *American Universities and Colleges*, sixth edition (American Council on Education, 1952), 54–57; eighth edition (American Council on Education, 1960), table 6.

23. Letter from I. S. Bowen to Jesse Greenstein, January 17, 1948, I. S. Bowen Papers, Huntington Library, box 10, folder 10.125.

24. Letter from I. S. Bowen to Vannevar Bush, April 10, 1952. I. S. Bowen Papers, Huntington Library, box 21, folder 21.341.

25. Warwick, *Masters of Theory*, esp. chapter 4. A decade earlier, Robert Kohler could only speculate "surely recruiting and teaching students must also affect the definition of research problems and modes of academic practice." See Kohler, "The PhD. Machine," *Isis* 81 (1990): 639–662, 658. In the wake of Warwick's wondrous work, a wealth of studies have appeared, including the following: David Kaiser, "'A Ψ is just a Ψ?'" *Studies in the History and Philosophy of Modern Physics*, 29 (1998), no. 3: 321–338; Suman Seth, "Crisis and the Construction of Modern Theoretical Physics," *British Journal for the History of Science* 40 (2007): 25–51; Catherine Jackson, "Visible Work," *Notes & Records of the Royal Society* 62 (2008): 31–49. Also Graeme Gooday, "Precision Measurement and the Genesis of Physics Teaching Laboratories in Victorian Britain," *British Journal for the History of Science* 23 (1990): 25–51.

26. McCray, *Giant Telescopes*; Ronald E. Doel, *Solar System Astronomy in America* (Cambridge University Press, 1996), 88 and 191–3; David DeVorkin, "Who Speaks for Astronomy?," *Historical Studies in the Physical and Biological Sciences* 31 (2000), no. 1: 52–92; Butrica, *To See the Unseen*; Smith, *The Space Telescope*; DeVorkin, *Science with a Vengeance*.

27. Kathryn M. Olesko, *Physics as a Calling* (Cornell University Press, 1991), 15.

28. See Robert Darnton, "Workers Revolt," in *The Great Cat Massacre and Other Adventures in French Cultural History* (Basic Books, 1984).

29. "The examination is the technique by which power, instead of emitting the signs of its potency, instead of imposing a mark on its subjects, holds them in a mechanism of objectification." Michel Foucault, *Discipline and Punish* (Penguin, 1991), 187.

30. George Chauncey, *Gay New York* (Basic Books, 1994). See also Jacqueline Fortes and Larissa Lomnitz, *Becoming a Scientist in Mexico* (Pennsylvania State University Press, 1994), esp. chapter 6; Simon Schwartzman, *A Space for Science* (Pennsylvania State University Press, 1991), 12–13.

31. David Kaiser, "Moving pedagogy from the periphery to the center," in *Pedagogy and the Practice of Science*, ed. D. Kaiser (MIT Press, 2005).

32. Stephen Hawking, "Introduction," in *From Newton to Hawking* (Cambridge University Press, 2003).

33. David Kaiser, "Cold War Requisitions, Scientific Manpower, and the Production of American Physicists after World War II," *Historical Studies in the Physical Sciences* 33 (2002), no. 1: 131–159. Also see Stuart W. Leslie, *The Cold War and American Science* (Columbia University Press, 1993), 10.

34. "Draft Proposal for Support of Astronomy by the National Science Foundation," n.d. (ca. 1951), Greenstein Papers, California Institute of Technology Archives, folder 27.2.

35. Most strongly noted by Krige (*American Hegemony and the Postwar Reconstruction of Science in Europe*, 2–3) and by Rana Mitter and Patrick Major ("Foreward," *Cold War History* 4, 2003, no. 1). It is the exceptions that are most interesting, esp. Traweek, *Beamtimes and Lifetimes*, and John Krige, "The Installation of High-Energy Accelerators in Britain after the War" in *The Restructuring of Physical Science in Europe and the United States, 1945–60* ed. M. De Maria (World Scientific, 1989).

36. I have to thank Hyungsub Choi for our discussions over this issue; Jonathan Harwood commented on "the dearth of cross-national analyses" in the history of science in the same year that Pnina Abir-Am first emphasized the international and transdisciplinary nature of molecular biology as a mode of scientific practice. Jonathan Harwood, *Styles of Scientific Thought* (University of Chicago Press, 1993), 3; Pnina Abir-Am, "From multidisciplinary collaboration to transnational objectivity: international space as constitutive of molecular biology, 1930–1970," in *Denationalizing Science*, ed. E. Crawford et al. (Kluwer, 1993). Scott Kirsch acknowledges a "far more active Soviet peaceful nuclear explosives program" that continued well beyond his American study of Project Plowshare, it warrants no comparative treatment, however cursory.

This point is the only weakness of Kirsch's book *Proving Grounds* (Rutgers University Press, 2005).

37. The Castro is a neighborhood of San Francisco that "like a small town evolved its own institutions and customs." Frances Fitzgerald, *Cities on a Hill* (Simon & Schuster, 1987), 20. Sugar Creek is a small rural town on the Illinois prairie. John Mack Faragher, *Sugar Creek* (Yale University Press, 1986); J. Ronald Engel, *Sacred Sand* (Wesleyan University Press, 1983).

38. *Communication with Extraterrestrial Intelligence (CETI)*, ed. C. Sagan (MIT Press, 1973), 2–3.

39. "Report from USA National Committee, Commission 5: Extra-terrestrial Radio Noise," in Donald Menzel, USRI Correspondence, 1946–1953, April, 1950, HUG4567.13, Harvard College Archives.

40. Handwritten letter from van de Hulst to Greenstein, April 26, 1951, Greenstein Papers, California Institute of Technology Archives, folder 41.1.

41. Minkowski's whole trip, traveling first class, was underwritten by the National Science Foundation's Panel on Radio Astronomy, which shared Minkowski's unhappiness with the data each group was contributing. See Letter from Tuve to Minkowski, November 30, 1955, Tuve Papers, Library of Congress, box 327, folder AUI-NSF Sept 1955→.

42. See "Notes made during a visit of some United States Radio Astronomy Observatories, March-Sept 1959," Oort Papers, University of Leiden library, box 175.

43. Paul Forman, "Scientific Internationalism and the Weimar Physicists," *Isis* 64 (1973): 151–180.

44. Ann Johnson, "Modeling Molecules," *Perspectives on Science* 17 (2009): 144–173, 147.

45. As Michael Gordin argues, chemists and physicists played major roles in the production of the fissionable materials uranium and plutonium. Michael D. Gordin, *Five Days in August* (Princeton University Press, 2007), 42. The "bat bomb" project sought to take refrigerated (and thus hibernating) bats, attach a small incendiary device to each, and release tubes full of bats over Japan. The plan was that the bats would wake up as they fell and would then roost in Japanese houses. After a period, the small "bat bombs" could be detonated, beginning small fires in a wide area. In one stranger anecdote, the leader of the "bat bomb" project, Dr. Lytle Adams, went to Washington in 1942 to seek appropriations and "some general" evidently "confused our secret program with another secret project" (the Manhattan Project) because they were both in the American southwest. Upon clarifying that he was dealing with bats and not atoms, he found that support from the Army was not forthcoming. Adams commented: "We got a sure thing like the bat bomb going, something that could really win the war, and they're jerking off with tiny little atoms. It makes me want to cry." Jack Couffer, *Bat Bomb* (University of Texas Press, 1992), 61. Thanks to Bruce Hunt for bringing this to my attention, and much else besides.

46. Can the hero worship of wartime scientists be better expressed than it is in the film *October Sky* (Universal Pictures, 1999)?

47. Louis Brown, *A Radar History of World War Two* (Taylor & Francis, 1999); Henry Guerlac, *RADAR in World War II* (Tomash/AIP, 1987); Robert Bud, *Penicillin* (Oxford University Press, 2007), esp. 63–75.

48. Steve Joshua Heims, *Constructing a Social Science for Postwar America* (MIT Press, 1993), 7.

49. Lloyd Berkner, *The Scientific Age* (Yale University Press, 1964), 59; Cohen, *Science, Servant of Man*, chapter 7.

50. David I. Kaiser, Making Theory, PhD thesis, Harvard University, 2000, 8.

51. Cathryn Carson, Ethan Pollock, Peter Westwick, and James H. Williams, "Editor's Foreword," *Historical Studies in the Physical and Biological Sciences* 30 (1999), no. 1, iii.

52. Stuart W. Leslie and Bruce Hevly, "Steeple Building at Stanford," *IEEE Proceedings* X (1985): 1161–1180.

53. Stuart W. Leslie, "Playing the Education Game to Win," *Historical Studies in the Physical and Biological Sciences* 18 (1987): 55–88, at 58.

54. Joel B. Hagen, *An Entangled Bank* (Rutgers University Press, 1992), 121. In fact India resisted building a bomb for many years, preferring to pursue atomic power for modernization, though by Indira Gandhi's time an atomic bomb had become a priority for the Indian state. See Itty Abraham, *The Making of the Indian Atomic Bomb* (Zed Books, 1998). Though heralded as ending the war and promising energy limited only by human imagination, the atomic bomb cast a pall over Cold War science and fundamentally "undermined the self-justification of the technological sublime, which had exulted the inventor and seen his material improvements as morally uplifting." David E. Nye, *American Technological Sublime* (MIT Press, 1999), 228.

55. James Carroll, *House of War* (Houghton Mifflin, 2006), 32.

56. Sir Robert Watson-Watt, *Man's Means to His End* (Heinemann, 1962), 133.

57. John Krige, *American Hegemony and the Postwar Reconstruction of Science in Europe* (MIT Press, 2006), 12.

58. See "United States Nuclear Tests," December 1994. DOE/NV-209 (Rev.14), viii. The year 1962 saw the most nuclear tests, 96 in all; those Democrats sure loved their explosions. My thanks to Bill Leslie and the History of Las Vegas class for the tour of the Nevada Test Site.

59. Peter Westwick, *The National Labs* (Harvard University Press, 2003), 12–13, 10, 178.

60. Michael Aaron Dennis, A Change of State, PhD thesis, Johns Hopkins University, 1990; Paul Forman, "Into Quantum Electronics," in *National Military Establishments and the Advancement of Science and Technology*, ed. P. Forman and J. Sánchez-Ron (Kluwer, 1996), 280; Joanne Goldman, "National Science in the Nation's Heartland," *Technology and Culture* 41 (2000): 432–459; Peter Westwick, *Into the Black* (Yale University Press, 2007).

61. Kirsch, *Proving Grounds*, 6.

62. Kirsch, *Proving Grounds*, 39; For another failure see, A. Kolb and L. Hoddeson, "The Mirage of the 'World Accelerator for World Peace' and the Origin of the SSC, 1953–1983," *Historical Studies in the Physical and Biological Sciences* 24 (1993), no. 1: 101–124.

63. See, for example, Odd Arne Westad, *The Global Cold War* (Cambridge University Press, 2007).

64. David Kaiser, "The Postwar Suburbanization of American Physics," *American Quarterly* 56 (2004): 851–888, at 873.

65. Michael Aaron Dennis, "Secrecy and Science Revisited," in *The Historiography of Contemporary Science, Technology, and Medicine*, ed. R. Doel and T. Söderqvist (Routledge, 2006), 172–184, at 175.

66. Michael Frayn, *Copenhagen* (Methuen Drama, 1998); Richard Rhodes, *The Making of the Atomic Bomb* (Penguin, 1988), chapters 5–6; Helge Kragh, *Dirac* (Cambridge University Press, 1990); *Rutherford and Physics at the Turn of the Century*, ed. M. Bunge and W. Shea (Dawson, 1979).

67. "Plan B," ACBL/EMH. 20 April 1944, Second World War Papers of Sir Bernard Lovell, Imperial War Museum, BL 7/1.

68. Letter from E. G. Bowen to Merle Tuve, March 16, 1955, AA C3830, A1/3/11/3 Pt. 3.

69. Merle Tuve to NSF Panel on Radio Astronomy, E. M. Purcell, Bart Bok, Jesse L. Greenstein, E. Minkowski, John Hagen, John D. Kraus, May 2, 1955, Merle Tuve Papers, Library of Congress, box 329, folder NSF Panel, Radio Astronomy, May-June 1955.

70. I. Bernard Cohen, *Science, Servant of Man* (Sigma Books, 1949), 141. Cohen drew this conclusion from case studies of nylon and rubber, radio and radar, and plants and organic herbicides.

71. Sir Edward Appleton, "Science for Its Own Sake," *Science* 119 (January 22, 1954): 103–109, 104, 109. In a parallel example, Richard Rhodes quotes Robert Oppenheimer as saying "It is a profound and necessary truth that the deep things in science are not found because they are useful; they are found because it was possible to find them." *The Making of the Atomic Bomb*, 11.

72. Appleton, "Science for Its Own Sake," 108.

73. Robert D. Putnam, *Bowling Alone* (Simon & Schuster, 2000), 24.

74. Mario Biagioli, Janet Browne, and Rob Iliffe have addressed this theme. Biagioli, *Galileo Courtier* (Chicago University Press, 1993); Browne, *Charles Darwin* (Jonathan Cape, 2002); Iliffe, "Material Doubts," *British Journal for the History of Science* 28 (1995): 285–318.

75. Bernard Lovell, *Astronomer by Chance* (Basic Books, 1990), 130.

76. Edge and Mulkay, *Astronomy Transformed*, 63.

77. Michael Aaron Dennis, "Accounting for Research," *Social Studies of Science* 17 (1987): 479–518, at 508.

78. Atsushi Akera, *Calculating a Natural World* (MIT Press, 2007), 2.

79. Sally G. Kohlstedt, *The Formation of the American Scientific Community* (University of Illinois Press, 1976); Daniel J. Kevles, *The Physicists* (Harvard University Press, 1995).

80. Kuhn, *The Structure of Scientific Revolutions*, 176, 184-185. While initially a "scientific community consists of men that share a paradigm" Kuhn concluded that the critical period of crisis is when "the members of a particular community must . . . choose between incompatible ways of practicing their discipline." In other words, any community is secondary to the demands of the discipline.

81. Johnson, "Modeling Molecules," 146, 147, 157. Johnson's use of "agendas in community formation," attributed to Michael Mahoney's study of computer science, seeks to unhinge incommensurability from Kuhn's theoretical paradigms in favor of a clash between competing agendas. Mahoney, however, is not talking about communities but about "the emergence of a discipline" and "what practitioners of a discipline agree," and what "becoming a recognized practitioner of a discipline means." See ibid., 149; Michael S. Mahoney, "Computer Science," in *Companion to Science in the Twentieth Century*, ed. J. Krige and D. Pestre (Routledge, 2003), at 619.

82. Elisabeth Crawford, Terry Shinn, and Sverker Sörlin, "The Nationalization and Denationalization of the Science," in *Denationalizing Science*, ed. E. Crawford et al. (Kluwer, 1993), at 15.

83. Kohler, *Lords of the Fly*; Rasmussen, *Picture Control*; Timothy Lenoir and Christophe Lécuyer, "Instrument Makers and Discipline Builders," *Perspectives on Science* 3 (1995): 276–345; Davis Baird, "Analytical Chemistry and the 'Big' Scientific Instrumentation Revolution," *Annals of Science* 50 (1993): 267–290; Kirsch, *Proving Grounds*, 3.

84. Doel, *Solar System Astronomy in America*, 88 and 191–193. In a telling aside, Sagan mentioned how "young scientists have again been attracted to planetary studies, not only astronomers, but also geologists, chemists, physicists, and biologists. The discipline needs them all." Carl Sagan, *The Cosmic Connection* (Papermac, 1981), 195. Sagan understood that a recent scientific discipline was an amalgamation of disciplines necessarily working cooperatively. I suggest that his understanding of a scientific "discipline" had become akin to community. And although David DeVorkin outlines a directly analogous case having to do with the impact of more general electronic techniques on astronomy, there were important differences. Astronomers themselves promoted the use of photoelectric cells, whereas external practitioners (notably electrical engineers and physicists) played a much larger role in the introduction of radio techniques into astronomy. See DeVorkin, "Electronics in Astronomy," 1207 and 1217. Allan Needell noted that even in the case of James van Allen, astronomers feared working with the military and especially held reservations about narrowing research and security restrictions. Allan A. Needell, "Preparing for the Space Age," *Historical Studies in the Physical and Biological Sciences* 18 (1987): 89–109, at 104–105.

85. Edge and Mulkay, *Astronomy Transformed*, 7. The case of Swedish radio astronomy also illustrates that many later stories now rely on Edge and Mulkay's basic formulation. While Mikael Hård looked at the technological transformation of German radar dishes into instruments of ionospheric and astronomical research, their introduction into an astronomical "observatory" is unproblematic, paralleling Edge and Mulkay's account. Mikael Hård, "Technological Drift in Science," in *Center on the Periphery*, ed. S. Lindqvist (Science History, 1993), 383–4. Among scientists own accounts too, authority stems from identifiable formation of a "discipline". At one commemorative occasion for radio astronomy, Bernard Burke of MIT said "it is a historical fact that radio astronomy started and grew as a technique-oriented discipline." Bernard Burke, "Radio Astronomy," in *Serendipitous Discoveries in Radio Astronomy*, ed. K. Kellermann and B. Sheets (NRAO/AUI, 1983). Benjamin Malphrus called radio astronomy "a new discipline" in his history of the United States' National Radio Astronomy Observatory. B. K. Malphrus, *The History of Radio Astronomy and the National Radio Astronomy Observatory* (West Virginia University, 1990). "This new discipline," claimed Graham Spinardi. Spinardi, "Science, technology, and the Cold War," *Cold War History* 6 (2006), no. 3: 279–300, at 280.

86. After emphasizing the role of independent journals and societies in forging disciplinary identity for new sciences, they confess: "Radio astronomy—unusually, for a new specialty—has no journal to give it professional identity." Edge and Mulkay, *Astronomy Transformed*, 64. A parallel case is in Johnson, "Modeling Molecules," 160, note 14.

87. "Remarks of Vannevar Bush at the 23rd Annual Scientific Assembly of the Medical Society of the District of Columbia. October 1, 1952, copy in I. S. Bowen Papers, Huntington Library, box 21, folder 21.342. Published as Vannevar Bush, "Science in Medicine and Related Fields," *Medical Annals of the District of Columbia* 22 (1953), no. 1: 1–7.

88. For a recent critique of the system, see George Weisz, *Divide and Conquer* (Oxford University Press, 2006), xi–x, 231ff.

89. David Hounshell, *From the American System to Mass Production, 1800–1932* (Johns Hopkins University Press, 1985), 229.

90. Jamie Cohen-Cole, "Instituting the Science of Mind," *British Journal for the History of Science* 40 (2007): 567–597, at 574.

91. Biochemistry evolved into the molecular biology of "interdisciplinary cooperation," says Lily E. Kay in *The molecular Vision of Life* (Oxford University Press, 1993), 7. Warren Weaver's vision for molecular biology was evidently a "transdisciplinary domain." Robert E. Kohler, *Partners in Science* (University of Chicago Press, 1991), 303. In the case of computational nanotechnology, Ann Johnson thought such labels merely the product of policy makers skirting the issue of who does nanotechnology. Yet Johnson too considers the specific "subfield" of computational nanotechnologists, notes how "nano does lie between traditional disciplines," notes that nano has a "substantial research program," concedes that nanotechnology "communities are usually highly multidisciplinary," and, finally, that nanotechnology is a "research field." Johnson, "Modeling Molecules," 145, 146, 147, 158.

92. For "discipline," see p. 3 of Edge and Mulkay, *Astronomy Transformed*; for "specialty," see p. 64; for "research community," see p. 6; for "branch," see p. 7; for "research networks," see p. 369; The problem of labeling the "motley" non-discipline of radio astronomy similarly plagued Woodruff Sullivan, who ended up calling the field a "research specialty." Sullivan, *Cosmic Noise*, 4.

93. This is indebted to the afterword of *Growing Explanations*, ed. M. Norton Wise (Duke University Press, 2004), and to the chapter in that volume by Claude Rosental ("Fuzzyfying the World"). Francis Crick admitted that he claimed the fuzzy descriptor "molecular biologist" because he "got tired of explaining that [he] was a mixture of crystallographer, biophysicist and geneticist." F. Crick, "Recent Research in Molecular Biology," *British Medical Bulletin* 21 (1963), no. 33: 183–186, at 184.

94. "Eighteenth Informal Memorandum from Harlow Shapley," August 8, 1952, UA V 630.28, B. J. Bok Administration files, box 1944-July 53, file June-Sept 1952, Harvard College Archives.

CHAPTER 1

1. Joan Freeman, *A Passion for Physics* (Adam Hilger, 1991), 94–95; Frank J. Kerr, "Early Days in Radio and Radar Astronomy in Australia," in *The Early Days of Radio Astronomy*, ed. W. Sullivan (Cambridge University Press, 1984), 133.

2. Bowen to Marsden, July 27, 1945, CRS C3830, A1/1/1, part 1. See also Marsden to Bowen, August 20, 1945, CRS C3830, A1/1/1, part 1. Ruby Payne-Scott, in a newspaper article to the *Sun*, confirmed that the first information on solar noise came from New Zealand. See "Solar Noise," October 8, 1946, CRS C3830, D9/2/2.

3. E. G. Bowen, "The Origins of Radio-Astronomy in Australia," in *The Early Years of Radio Astronomy*, ed. W. Sullivan (Cambridge University Press, 1984), 87–88.

4. Christiansen and Mills, *Dr. J.L. Pawsey: Obituary*, 1962, Pawsey Papers, Basser Library, Australian Academy of Science, MS 20.

5. B. Lovell, "Joseph Lade Pawsey," *Biographical Memoirs of Fellows of the Royal Society* 10 (1964): 230–231.

6. Lovell, "Joseph Lade Pawsey," 234.

7. J. P. Wild, "The Beginnings of Radio Astronomy in Australia," *Records of the Australian Academy of Science* 2 (1972), no. 3: 53; W. Miller Goss and R. X. McGee, *Under the Radar* (Springer, 2009).

8. J. Pawsey, R. Payne-Scott, and L. McCready, "Radio-Frequency Energy from the Sun," *Nature* 157 (1946): 158–159.

9. Gale E. Christianson, *Edwin Hubble* (Farrar, Straus and Giroux, 1995).

10. Daniel J. Kevles, *The Physicists* (Harvard University Press, 1995), 355.

11. Goss and McGee, *Under the Radar*, 5–6.

12. Edwin Hubble to Vannevar Bush, September 10, 1945, Edwin Hubble Papers, Huntington Library, box 13, folder HUB414.

13. Charles A. Whitney, *The Discovery of our Galaxy* (Knopf, 1971), 241.

14. Hubble to Bowen, October 16, 1945, and letter from Bush to Bowen, October 20, 1945, I. S. Bowen Papers, Huntington Library, box 1, folder 1.6. Ira Bowen was appointed director of Mount Wilson in 1946 and was given the dual directorship upon the amalgamation of the institutions in 1948. See H. W. Babcock, "Ira Sprague Bowen," *Biographical Memoirs, National Academy of Sciences* 53 (1982): 82–119, 109; Donald Osterbrock, "The Appointment of a Physicist as Director to the Astronomical Center of the World," *Journal for the History of Astronomy* 23 (1992): 155–165.

15. Custer Baum to I. S. Bowen, July 1946. I. S. Bowen Papers, Huntington Library, box 2, folder 2.10.

16. David Kaiser, "The Postwar Suburbanization of Physics," *American Quarterly* 56 (2004): 851–888, at 870.

17. Bowen to Bush, November 13, 1945, I. S. Bowen Papers, Huntington Library, box 1, folder 1.6.

18. Bush to Hubble, October 3, 1946, I. S. Bowen Papers, Huntington Library, box 1, folder 1.6. See the entire folder for the cautious treatment of Hubble by Bowen, Bush, and Max Mason (chairman of the search committee) leading up to the October announcement.

19. CIW, *Annual Report 1945–46*, 1–3, I. S. Bowen Papers, Huntington Library, box 1, folder 1.1. See also Judith R. Goodstein, *Millikan's School* (Norton, 1991), 275.

20. See David Edgerton, *Warfare State* (Cambridge University Press, 2006), 166–181, esp. 176–77; *Cold War, Hot Science*, ed. R, Bud and P. Gummett (Science Museum, 1999), chapters 1, 2, 11, and (esp.) 8.

21. Bernard Lovell, *Echoes of War* (Adam Hilger, 1991).

22. Bernard Lovell, *Astronomer by Chance* (Basic Books, 1990), 109.

23. *A Centre of Intelligence*, ed. C. Field and J. Pickstone (Johns Rylands Library, 1988); Mary Croarken, "The Beginnings of the Manchester Computer Phenomenon," *IEEE Annals of the History of Computing* 15 (1993), no. 3: 9–16; Robert H. Kargon, *Science in Victorian Manchester* (Manchester University Press, 1977).

24. Lovell to Blackett, July 1, 1945, John Rylands Library/Jodrell Bank Archives, CS7, box 19, file 4.

25. Lovell to Ratcliffe, July 29, 1945, John Rylands Library/Jodrell Bank Archives, CS7, box 19, file 4.

26. John Rylands Library/Jodrell Bank Archives, CS7, box 19, file 4.

27. ACBL/RMY, May 19, 1944, Second World War Papers of Sir Bernard Lovell, Imperial War Museum, BL 7/1.

28. M.P. Applebey, ICI, to Ryle, April 5, 1945. Ryle Papers, Churchill College, Cambridge University, box 2, folder 4.

29. Ryle Senior to Rowe, May 24, 1945, plus handwritten note to Martin. Ryle Papers, Churchill College, Cambridge University, box 2, folder 4.

30. Ryle to Plaskett, March 15, 1952. Ryle Papers, Churchill College, Cambridge University, box 2, folder 4.

31. Marc Rothenberg, "Organization and Control," *Social Studies of Science* 11 (1981): 305–325; Grote Reber and Jesse Greenstein, "Radio-frequency Investigations of Astronomical Interest," *The Observatory* 67 (1947): 15–26, 15.

32. Jesse Greenstein and F. L. Whipple, "On the Origin of the Interstellar Radio Disturbances," *Proceedings of the National Academy of Sciences* 23 (1937): 177.

33. Greenstein, *Sources for History of Modern Astrophysics*, 72. See J. Greenstein, L. Henyey, and P. G. Keenan, "Interstellar Origin of Cosmic Radiation at Radio-Frequencies," *Nature* 157 (1946): 805–806.

34. Reber to Greenstein, November 19, 1946, Greenstein Papers, California Institute of Technology Archives, folder 31.16. Approaching Harvard, Reber was acknowledged but essentially ignored, and similar negotiations at the McDonald Observatory, through ONR, and with Edward Condon at the National Bureau of Standards both failed as Reber demanded a "large salary" from a personal research contract with the institutions concerned. Greenstein estimated that Reber was tripling the research costs. Letter from Greenstein to Otto Struve, November 27, 1946, Greenstein Papers, California Institute of Technology Archives, 36.17. Eventually, though still only partially, Reber would fulfill his dreams in Tasmania, Australia, during in the early 1960s. See entire file for continuous news from Reber about failed sponsorship endeavors. See also Jesse Greenstein, Oral History interview with Spencer Weart, April 7, 1977. *Sources for History of Modern Astrophysics,* 72.

35. Minutes of Meeting of the Observatory Council, October 8, 1946, UA V 630.28.5, Harvard College Archives.

36. Robinson to Lovell, November 15, 1945, John Rylands Library/Jodrell Bank Archives, CS6, box 1, file 1.

37. Freeman, *A Passion for Physics.*

38. P. M. S. Blackett, *Fear, War and the Bomb* (McGraw-Hill, 1948), 106.

39. Abraham, *The Making of the Indian Atomic Bomb.*

40. James Phinney Baxter, *Scientists Against Time* (MIT Press, 1968). The last chapter is on the atomic bomb, though the material is almost completely taken from the Smyth Report.

41. Notes for John Briton from a conversation with Prof. L. H. Martin, n.d. (ca. early 1945.), CRS C3830, E2/2 Pt. 1.

42. Bowen to Rivett, April 3, 1945, CRS C3830, A2/1, part 2.

43. Alice Cawte, *Atomic Australia* (University of New South Wales Press, 1992), 7.

44. Luis W. Alvarez, *Alvarez* (Basic Books, 1987), 154.

45. Robert Seidel, "Accelerating Science," *Historical Studies in the Physical Sciences* 13 (1983): 375–400, at 380.

46. In 1942, Otto Struve, then director of the Yerkes Observatory, commented that he was "glad that it is possible for the department to take and active part in the training of men," through instruction in navigation. Moreover, he was relieved that the course made "it a little more convincing that members of the Observatory staff do fill positions of importance to the conduct of the war." Letter from O. Struve to Greenstein, May 29, 1942, Greenstein Papers, California Institute of Technology Archives, 36.17.

47. Lovell, *Echoes of War*, 17.

48. See Mary Jo Nye, *Blackett* (Harvard University Press, 2004), 76.

49. Alvarez, *Alvarez*, 88.

50. Both Bowen and Rowe mention friction between themselves after Rowe's appointment in their memoirs. E. G. Bowen, *Radar Days*. (Adam Hilger, 1987), 50; A. P. Rowe, *One Story of Radar*. (Cambridge University Press, 1948), 25; see also R. Hanbury Brown, H. C. Minnett, and F. W. G. White, "Edward George Bowen," *Biographical Memoirs of Fellows of the Royal Society* 38 (1992): 43–65, 50–51. In a further twist, Rowe did not long survive the postwar shakeup of British science; he went to Australia, where he was vice-chancellor of the University of Adelaide from 1948 to 1955. See Edgerton, *Warfare State*, 167. I know of no correspondence between Bowen and Rowe, though they both moved to Australia.

51. Alan Waterman of the NSF acknowledged that the term "basic research" itself only received general recognition with Bush's *Endless Frontier*. Within the context of Waterman's article it is clear that general recognition implies American recognition. See Alan Waterman, "Basic Research in the United States," in *Symposium on Basic Research* (American Association for the Advancement of Science, 1959), 20.

52. Vannevar Bush, *Science—The Endless Frontier* (US Government Printing Office, 1945), 2–3; Daniel J. Kevles, "Principles and Politics in Federal R&D Policy, 1945–90," preface to *Science— The Endless Frontier* (National Science Foundation, 1990), ix; Daniel Kleinman, *Politics on the Endless Frontier* (Duke University Press, 1995).

53. Copy of Vannevar Bush's address before the General Engineering Meeting, Washington, D.C., "Government and Research," January 14, 1947, I. S. Bowen Papers, Huntington Library, box 4, folder 4.50.

54. David M. Hart, *Forged Consensus* (Princeton University Press, 1998), 184–85.

55. Paul Forman, "Into Quantum Electronics," in *National Military Establishments and the Advancement of Science and Technology*, ed. P. Forman and J. Sánchez-Ron (Kluwer, 1996), at 265.

56. An exception is chapter 1 of Matthew H. Wisnioski, *Engineers for Change* (MIT Press, 2012), which discusses Bush's similar ideas about engineering.

57. G. Pascal Zachary, *Endless Frontier* (Free Press, 1997), 219. This interpretation is at odds with Daniel Kevles' view of Bush's "prewar experience," particularly that with the National Advisory Committee for Aeronautics, as central to the emergence of the *Endless Frontier* report. See Kevles, "Principles and Politics in Federal R&D Policy, 1945–90," x.

58. Bush, *Science*, 4.

59. Oliver D. Hensley, "Traditional Models for Classifying University Research," in *The Classification of Research*, ed. O. Hensley (Texas Tech University Press, 1988), 7.

60. Goodstein, *Millikan's School*, 269.

61. Stuart W. Leslie, *The Cold War and American Science* (Columbia University Press, 1993); Rebecca Lowen, *Creating the Cold War University* (University of California Press, 1997).

62. Michael Aaron Dennis, A Change of State, Ph.D. dissertation, Johns Hopkins University, 1990.

63. David Kaiser, Making Theory, Ph.D. thesis, Harvard University, 2000, 73.

64. Ibid., 45.

65. Geoffrey Bolton, *The Oxford History of Australia*, volume 5 (Oxford University Press, 1996), 28–30.

66. C. Boris Schedvin, *Shaping Science and Industry* (Allen & Unwin, 1987), 310–311.

67. T. Rowse, *Australian Liberalism and National Character* (Kibble Books, 1978), 132.

68. H. C. Coombs, *Trial Balance* (Macmillan, 1981), 26.

69. Schedvin, *Shaping Science and Industry*, 310.

70. Ibid, 323–324.

71. Ibid., 323. Quote from Dedman in 1943.

72. David Rivett, *Science and Responsibility*, Rivett Papers, Basser Library, Australian Academy of Science, MS 83/9, a.

73. Nicolas Brown, *Governing Prosperity* (Cambridge University Press, 1995), 210.

74. Bush to Bowen, November 23, 1945, I. S. Bowen Papers, Huntington Library, box 1, folder 1.6. Bowen agreed with Bush's remark about "too much emphasis" on nuclear physics; see letter from Bowen to Bush, December 7, 1945, I. S. Bowen Papers, Huntington Library, box 1, folder 1.6.

75. For example, Correlli Barnett, *The Lost Victory* (Macmillan, 1995).

76. Schedvin, *Shaping Science and Industry*, 312.

77. F. W. G. White, *A Discussion of CSIR*, July 19, 1944, CRS C3823, E12/2.

78. H. C. Minnett and Sir Rutherford Robertson, "Frederick William George White," *Historical Records of Australian Science* 11 (1996), no. 2: 239–258; David Ellyard, interview with F. W. G. White, November 3, 1979, CSIRO Archives, series 528.

79. Lens making was undertaken at various observatories. Valve manufacture was instigated at the University of Melbourne, under Laby and L. H. Martin, before being given over to Amalgamated Wireless Valve Company. See D. E. Caro and R. L. Martin, "Leslie Harold Martin," *Historical Records of Australian Science* 7 (1987), no. 1: 97–107, 99–100.

80. Report to the Executive on postwar reconstruction, March 22, 1943, CRS C3823, E12/2; D. T. Gillespie, "Research Management in the Commonwealth Scientific and Industrial Research Organization, Australia," *Public Administration (London)* 46 (1964): 11–31, 25.

CHAPTER 2

1. G. Julius, "The growth of science in Australia," *Sydney Morning Herald*, December 29, 1945, CRS C3830, D9/4 (a); X. Primrose, A Powerful Paradox, B.A. thesis, Australian National University, 1994; Alice Cawte, *Atomic Australia, 1944–1990* (University of New South Wales Press, 1992).

2. Bowen to White, August 9, 1945, CRS C3830, A2/1, part 2, "Uranium Bombs." Bowen offered Martin, Pawsey, Burhop, Hill, Gooden, and Allen as the Australian team.

3. Radiophysics received a single copy of the Smyth Report. J. L. Pawsey, "Atomic Power and American Work on the Development of the Atomic Bomb," *Australian Journal of Science* 8 (August 1945): 41–47.

4. Bowen to White, September 13, 1945, CRS C3830, E2/2, part 1; E. G. Bowen, O. O. Pulley and J. S. Gooden, "Application of Pulse Technique to the Acceleration of Elementary Particles," *Nature* 157 (1946): 840; E. L. Ginzton, W. W. Hansen, and W. R. Kennedy, "A Linear Electron Accelerator," *Review of Scientific Instruments* 19 (1948): 89–108.

5. CSIR, *20th Annual Report. Year ended 30th June 1946*, 89.

6. Seidel, "Accelerating Science," 389.

7. Peter Galison, Bruce Hevly and Rebecca Lowen, "Controlling the Monster," in *Big Science*, ed. P. Galison and B. Hevly (Stanford University Press, 1992), 58–59.

8. John Heilbron and Robert Seidel, *Lawrence and His Laboratory* (University of California Press, 1989), 46–47.

9. F. G. W. White to E. G. Bowen, July 5, 1946, CRS C3830, A2/5A/1. See also David P. D. Munns, "Linear Accelerators, Radio Astronomy, and Australia's Search for International Prestige, 1944–1948," *Historical Studies in the Physical and Biological Sciences* 27 (1997), no. 2: 299–317.

10. See CSIR annual reports for June 20, 1946 and June 22, 1948. On continuing efforts in particle acceleration, see R. W. Home, "The Rush to Accelerate," *Historical Studies in the Physical and Biological Sciences* 36 (2006), no. 2: 213–241.

11. CSIR, *21st Annual Report. Year ended 30th June 1947*, 99.

12. Bernard Lovell, *Astronomer by Chance* (Basic Books, 1990), 120.

13. A deck chair is one of the pieces in the special edition of the board game Astronomy Monopoly, alongside with a radio telescope, binoculars, an observatory, and the Hubble Space Telescope.

14. Lovell, *Astronomer by Chance*, 130.

15. Lovell to Bowen, September 3, 1946, John Rylands Library/Jodrell Bank Archives, CS6, box 1, file 1.

16. Lovell to Brearley, March 1947, John Rylands Library/Jodrell Bank Archives, CS6, box 1, file 1.

17. M. Ryle to Pawsey, March 21, 1950, CRS C3830, A1/1/1, part 5.

18. Pawsey to Ryle, March 31, 1950, CRS C3830, A1/1/1, part 5.

19. Pawsey to Ratcliffe, November 3, 1950, CRS C3830, A1/1/1, part 5.

20. Coubro and Scrutton to Lovell, August 10 1949, John Rylands Library/Jodrell Bank Archives, ACC, box 7, file 4.

21. John Rylands Library/Jodrell Bank Archives, ACC, box 13, file 10. The artisan Robert Hooke in his coffeehouse networks offers a parallel case. See Rob Iliffe, "Material Doubts," *British Journal for the History of Science* 28 (1995): 285–318.

22. Draft Appendix, ca. 1947, John Rylands Library/Jodrell Bank Archives, ACC, box 57, file 6. Frank Moran was on a Further Education grant; the other graduate students J. G. Davies, W. A. Hughes, I. A. Gatenby, and A. Aspinal were supported by DSIR funding.

23. Lovell and Ratcliffe, "Report on PhD thesis of C. Ellyett," December 4, 1948, John Rylands Library/Jodrell Bank Archives, CS5, box 2, file 2.

24. "PhD Examination," June 1, 1949, John Rylands Library/Jodrell Bank Archives, CS5, box 2, file 2.

25. Lovell to Plaskett, March 6, 1952, John Rylands Library/Jodrell Bank Archives, CS5, box 2, file 2.

26. Plaskett to Lovell, March 10, 1952, John Rylands Library/Jodrell Bank Archives, CS5, box 2, file 2.

27. Lovell to Findlay-Freundlich, November 7, 1952, John Rylands Library/Jodrell Bank Archives, CS5, box 2, file 2.

28. Part 2 Examination: Radio Astronomy, John Rylands Library/Jodrell Bank Archives, CS3, box 27, file 2, n.d. (1951).

29. Ronald Doel, for instance, does not identify any "curriculum change" in astronomical instruction "through the mid-1950s" (*Solar System Astronomy*, 196). I argue that such a view delays the admission of the new techniques into a discipline until fully certified professionals occupy established positions.

30. "Preliminary Examination for the Doctor's Degree," February 20, 1954, UA V 630.28, box "Bok, Administrative Files, August 1953–January 1957," folder Jan–March 1954, HCA.

31. Greenstein to P. Van de Kamp, June 14, 1954, Greenstein Papers, California Institute of Technology Archives, 112.3.

32. Greenstein to Seeger, November 5, 1952, California Institute of Technology Archives, 112.1.

33. CIW, *Annual Report 1945–46*, 1–3, I. S. Bowen Papers, Huntington Library, box 1, folder 1.1.

34. Greenstein to van de Hulst, April 11, 1950, Greenstein Papers, California Institute of Technology, folder 41.1. Van de Hulst was in total sympathy with Caltech's position on cooperative research. Writing a year later from Harvard, van de Hulst remarked on the more "interesting" and "pleasant contacts with the physics dept and with some MIT people." Handwritten letter from van de Hulst to Greenstein, April 26, 1951, Greenstein Papers, California Institute of Technology Archives, 41.1.

35. DuBridge to Shapley, February 25, 1947, DuBridge Papers, California Institute of Technology Archives, 34.9.

36. Copy of Greenstein's letter to Dean E. C. Watson, December 14, 1947, I. S. Bowen Papers, Huntington Library, box 5, folder 5.64.

37. Greenstein, *Sources for History of Modern Astrophysics*, 116.

38. Harlow Shapley to DuBridge, February 20, 1947, DuBridge Papers, California Institute of Technology Archives, 34.9.

39. DuBridge to Shapley, February 25, 1947, DuBridge Papers, California Institute of Technology Archives, 34.9.

40. Harlow Shapley to DuBridge, February 20, 1947, DuBridge Papers, California Institute of Technology Archives, 34.9.

41. Copy of Watson's letter to Greenstein, December 14, 1947, I. S. Bowen Papers, Huntington Library, box 5, folder 5.64.

42. Greenstein to Bowen, January 18, 1948, I. S. Bowen Papers, Huntington Library, box 10, folder 10.125.

43. Greenstein to Bok, September 29, 1948, Greenstein Papers, California Institute of Technology Archives, 3.25.

44. Greenstein, *Sources for History of Modern Astrophysics*, 91, 111.

45. Greenstein to Bok, September 29, 1948, Greenstein Papers, California Institute of Technology Archives, 3.25. Greenstein noted that many of the students attending his classes came from physics. See Greenstein to Strömgren, November 9, 1948, I. S. Bowen Papers, Huntington Library, box 10, folder 10.125.

46. Rudolph Minkowski taught Donald Osterbrock similarly. See McCray, *Giant Telescopes*, 29.

47. Greenstein, *Sources for History of Modern Astrophysics*, 113–114.

48. Dennis Overbye, *Lonely Hearts of the Cosmo* (HarperCollins, 1991), esp. chapter 1.

49. Draft: *Australian Aviation Annual*, December 17, 1948, CRS C3830, D8.

50. E. G. Bowen, "Programme of the Division of Radiophysics: 1946," August 8, 1946, CSIRO Archives, series 4, C21/6.

51. White to Munro & Gresford (S.R.L.O., London), August 11, 1944, CRS C3823, E7/1/2., 4.

52. Bowen to White, October 21, 1946, CRS C3830, E2/2, part 1.

53. Bowen to White, May 31, 1946, CSIRO Archives, series 9, G23/3.

54. Bowen, *Radar Days*, 201.

55. E. G. Bowen, "Programme of the Division of Radiophysics: 1946," 4. August 8, 1946, CSIRO Archives, series 4, C21/6.

56. J. G. Bolton, "Discrete Sources of Galactic Radio Frequency Noise," *Nature* 162 (July 1948): 141–142.

57. J. P. Wild, "The Beginnings of Radio Astronomy in Australia," *Records of the Australian Academy of Science* 2 (1972), no. 3: 54; Frank Kerr, "Early Days in Radio and Radar Astronomy in Australia," in *The Early Years of Radio Astronomy*, ed. W. Sullivan III (Cambridge University Press, 1984), 134.

58. Bolton Interview, Australian National Library, TRC324, tape 2.

59. Pawsey, "Solar and cosmic noise research in the United States and Canada," ca. 1948–49, CRS C3830, A1/1/5, part 2.

60. John G. Bolton, "History of Australian Astronomy," *Proceedings of the Astronomical Society of Australia* 4 (1982), no. 4: 349–358, at 352.

61. Pawsey, "Solar and Cosmic Noise Research in the United States and Canada," ca. 1948, CRS C3820, A1/1/5, part 2.

62. Submission from Bowen to Cook, March 15, 1950, CSIRO Archives, series 3, PH/BOL/5B.

63. Greenstein to Dr. R. v.d. R. Woolley, Commonwealth Observatory, Mount Stromlo, Australia, June 18, 1952, Greenstein Papers, California Institute of Technology Archives, 115.1.

64. Greenstein to van de Hulst, May 12, 1950, Greenstein Papers, California Institute of Technology, folder 41.1. David Kaiser points out that by the late 1950s Luis Alvarez was complaining that postwar graduate students lacked intellectual curiosity, a feature of the "suburbanization" of physics. David Kaiser, "The Postwar Suburbanization of American Physics," *American Quarterly* 56 (2004): 851–888, at 868.

CHAPTER 3

1. Bart Bok to Jesse Greenstein, November 26, 1954, Greenstein Papers, California Institute of Technology Archives, 3.26.

2. Ronald Clark, *The Rise of the Boffins* (Phoenix House, 1962); Francis Spufford, *Backroom Boys* (Faber & Faber, 2003). Robert Watson-Watt himself defined the boffin as "the instrument for building into the design provisions [for] field conditions in which the device is to operate, and above all things, the competence of those who are to operate, maintain, and repair it." R. Watson-Watt, *Three Steps to Victory* (Odhams, 1957), 255.

3. Hanbury Brown, *Boffin*, 96.

4. Andrew Warwick, *Masters of Theory* (University of Chicago Press, 2003), 479ff.

5. Bernard Lovell, *Astronomer by Chance* (Macmillan, 1990), 155.

6. Hanbury Brown, *Boffin*, 97. Also see A. C. B. Lovell, "Robert Hanbury Brown," in *Modern Technology and its Influence on Astronomy*, ed. J. Wall and A. Boksenberg (Cambridge University Press, 1990).

7. Hanbury Brown, *Boffin*, 103.

8. Zdenêk Kopal, *Of Stars and Me* (Adam Hilger, 1986), 245.

9. Lovell, "Confidential," May 30, 1949, John Rylands Library/Jodrell Bank Archives, CS3, box 31, file 2.

10. Ibid.

11. Bruno Latour and Steve Woolgar, *Laboratory Life* (Princeton University Press, 1986), 36.

12. "Preparations for R.A.S Visit," June 13, 1949, John Rylands Library/Jodrell Bank Archives, CS3, box 31, file 2.

13. Woody Sullivan generalized a claim that the new visual culture of radio astronomy evoked the radio astronomer's desire to become part of the discipline of astronomy. Sullivan, *Cosmic Noise*, 15.

14. Typed remarks on papers at the Royal Astronomical Society meeting in Manchester, n.d., John Rylands Library/Jodrell Bank Archives, CS3, box 31, file 2.

15. Secretary of the Royal Astronomical Society to Lovell, July 5, 1949, John Rylands Library/ Jodrell Bank Archives, CS3, box 31, file 2.

16. W. M. L. Greaves, Royal Observatory, Edinburgh, to Ryle, January 24, 1950, Ryle Papers, Churchill College, Cambridge University, box 1, folder 1.

17. "Minutes of the 1st Meeting of the 'Committee for Radio-Astronomy,'" February 10, 1950, Ryle Papers, Churchill College, Cambridge University, box 7, folder 15.

18. "Recommendation by The Council of the Royal Astronomical Society," June 1950, John Rylands Library/Jodrell Bank Archives, ACC, box 57, file 6.

19. DSIR to C. A. Spenser, Manchester Vice-Chancellor, July 2, 1948, John Rylands Library/ Jodrell Bank Archives, ACC, box 57, file 6.

20. Bowen, "The Origins of Radio-Astronomy in Australia," 92–93; Peter Robertson, "John Bolton and Australian Astronomy," *Australian Physicist* 21 (1984), no. 8: 178–180, at 178.

21. Bowen to Pawsey, December 24, 1947, AA C3830, F1/4/PAW/1, part 1; Richard v. d. R. Woolley, "Galactic Noise," *Monthly Notices of the Royal Astronomical Society* 107 (1947): 308–315. The paper described some of the theoretical considerations of free-free transitions in hydrogen that might generate radio noise, but didn't acknowledge any Radiophysics Laboratory work.

22. Woolley once noted to Bowen that the Royal Astronomer had mentioned Australia's progress in "this branch of physics [noise research]." Woolley to Bowen, April 14, 1947, CSIRO Archives, series 3, KE20/1.

23. "Routine Observations in Radio Astronomy at Cambridge," February 18, 1950, Ryle Papers, Churchill College, Cambridge University, box 1, folder 1.

24. Martin Ryle and Anthony Hewish, "The Effects of the Terrestrial Ionosphere on the Radio Waves from Discrete Sources in the Galaxy," *Monthly Notices of the Royal Astronomical Society* 110 (1950): 381. Cited in Edge and Mulkay, *Astronomy Transformed*, 30.

25. The myth appears especially strong on pp. 430–432 of Sullivan, *Cosmic Noise*.

26. Robert W. Smith, "Beyond the Big Galaxy," *Journal for the History of Astronomy* 37 (2006): 307–342, at 335–336.

27. F. Graham-Smith to Walter Baade, August 22, 1951. For Baade's reply, see Baade to Smith, September 3, 1951, Walter Baade Papers, Huntington Library, box 19, folder 37.

28. Dorrit Hoffleit, "New Notes," *Sky and Telescope* 10 (November 1951): 10.

29. Baade to Smith, November 25, 1951, Walter Baade Papers, Huntington Library, box 19, folder 37.

30. Baade to Ryle, November 26, 1954, Ryle Papers, Churchill College, Cambridge University, box 1, folder 3.

31. Smith to Baade, December 14, 1951, and Baade to Lovell, May 26, 1952, Walter Baade Papers, Huntington Library, box 19, folder 37.

32. Peter Galison, *Image and Logic* (University of Chicago Press, 1997), 782.

33. Martin Ryle, "Radio Astronomy," in *Search and Research*, ed. J. Wilson (Mullard Limited, 1971), 23.

34. Sullivan, *Cosmic Noise*, 165.

35. "The Third-Year Honours Physics Course Session 1948-49," John Rylands Library/Jodrell Bank Archives, CS3, box 28, file 3.

36. Pawsey to Mark Oliphant, September 10, 1951, CRS C3830, A1/1/4.

37. E. G. Bowen to Oort, January 31, 1952, Jan Oort Papers, University of Leiden Library, box 105, folder a. The plan was that Hill would return to the Radiophysics Laboratory after Leiden. See Bowen to Oort, September 11, 1952. For Pawsey's comment on the exceptional status of Hill, see Pawsey to Oort, March 14, 1955, Jan Oort Papers, University of Leiden Library, box 109, folder d.

38. Pawsey to Oort, September 11, 1952, Jan Oort Papers, University of Leiden Library, box 105, folder c.

39. Pawsey to Oort, March 14, 1955, Jan Oort Papers, University of Leiden Library, box 109, folder d.

40. Oort to Pawsey, October 30, 1952, Oort Papers, University of Leiden Library, box 105, folder c.

41. "Postgraduate Research Studentships Tenable in Australia," n.d., JRL/JBA, ACC, box 55, file 3.

42. Ryle to Ratcliffe, May 21, 1951, Ryle Papers, Churchill College, Cambridge University, box 1, folder 2.

43. Ryle to Lovell, n.d. (ca. 1951), Ryle Papers, Churchill College, Cambridge University, box 1, folder 2.

44. Ryle to Ratcliffe, June 21, 1951, Ryle Papers, Churchill College, Cambridge University, box 1, folder 2.

45. Edge and Mulkay, *Astronomy Transformed*, 23.

46. N. C. Mullins, "The Development of a Scientific Specialty," *Minerva* 10 (1972): 51–82.

47. Edge and Mulkay, *Astronomy Transformed*, 22–23.

48. Ryle to Ratcliffe, June 21, 1951, Ryle Papers, Churchill College, Cambridge University, box 1, folder 2.

49. Edward Appleton, "IXth General Assembly of U.R.S.I.," September 1950, John Rylands Library/Jodrell Bank Archives, CS3, box 31, file 4.

50. Woodruff T. Sullivan III, "Kapteyn's Influence on the Style and Content of Twentieth Century Dutch Astronomy," in *The Legacy of J. C. Kapteyn* (Kluwer, 2000), 245–251. This section is also indebted to chapter 16 of Sullivan, *Cosmic Noise*.

51. Reber to Greenstein, November 19, 1946, Greenstein Papers, California Institute of Technology Archives, 31.16. In the fickle memory of Reber the outsider, it was Reber himself who told van de Hulst that detection of the line was beyond "present technological capabilities." The comment from an interview in 1975 seems to defensively remember that period when Reber and Greenstein had published on the 21-centimeter line, but Reber's own efforts came to little. Sullivan, *Cosmic Noise*, 396.

52. Greenstein to Oort, July 11, 1951, Greenstein Papers, California Institute of Technology Archives, 27.7.

53. "American Astronomers Report," *Sky and Telescope* 10 (August 1951), 237–238.

54. Harold I. Ewen, Radiation from Galactic Hydrogen at 1420 MC/sec, PhD thesis, Harvard University, 1951, 1–2, Harvard College Archives, HU 90.5951.

55. Robert Buderi, *The Invention That Changed the World* (Simon & Schuster, 1996), 291–292, 295–304.

56. Apart from Ewen and Purcell, only the Bell Telephone Laboratories constructed a horn-type antenna for radio astronomy work. That antenna, like the HCO's, was ideal for investigating specific questions but was not a general survey instrument. With "calculable aperture efficiency," horn antennas restrict wavelength detection to the particular design specifications. J. S. Hey, *The Evolution of Radio Astronomy* (Science History Publications, 1973), 98–99.

57. Sullivan, "Kapteyn's Influence," 244.

58. Harold I. Ewen, Radiation from Galactic Hydrogen at 1420 MC/sec, PhD thesis, Harvard University, 1951, 1–2, Harvard College Archives, HU 90.5951.

59. Buderi, *The Invention That Changed the World*, 305.

60. Sullivan, "Kapteyn's Influence," 229, 261.

61. Bessie Jones and Lyle Boyd, *The Harvard College Observatory* (Belknap, 1971), 267.

62. Peggy Kidwell, "E. C. Pickering, Lydia Hinchman, Harlow Shapley, and the Beginning of Graduate Work at the Harvard College Observatory," *Astronomy Quarterly* 5 (1986): 157–171,

at 162. Also Morton and Phyllis Keller, *Making Harvard Modern* (Oxford University Press, 2001), 107.

63. See American Council on Education, *American Universities and Colleges*. This was the estimate of Donald Menzel, the HCO's director from 1952to 1966. See Donald Menzel, unpublished autobiography, ca. 1974, Harvard College Archives, HUG4567.3.

64. Aside from her own autobiography, Cecilia Payne-Gaposchkin has received little historical attention. Peggy Kidwell's exploratory chapter playing with the metaphor of "family" to elucidate some of the struggles faced by a woman scientist failed to reveal much about her subject's science and presented little significant historical detail. Indeed, the chapter revealed far more about the impoverished state of women's writing in the history of science than about the career of Payne-Gaposchkin. Seeking to uncover connections between marriage, family, children, and career (subjects often entirely overlooked in discussions of male scientists), instead Payne-Gaposchkin's biography looks like a romance novel, with more attention paid to an unnamed suitor than to her significant dissertation. Though marriage was undoubtedly important for many, Peggy Kidwell can pass over Arthur Eddington's bachelorhood without comment, as Matthew Stanley's recent work does as well, while dwelling at length on the spinsterhood that was the condition of many women scientists. Payne-Gaposchkin seems to have succeeded in spite of her husband because she relegated much of the domestic sphere to him. At least in the case of Payne-Gaposchkin, the co-construction of a woman and a scientist appears: while men can be unassuming bachelors devoted to science, women can be successful scientists but must also be successful women by marrying, bearing children, and maintaining a home. Such a double standard is both morally wrong and historically invalid, but also misses a larger issue. Kidwell missed an opportunity to reveal the culture of an observatory family and the changes that took place at the Harvard Observatory, in particular once Harlow Shapley retired from his fatherly role. Investigating the impact of such an observatory culture on all of its members, whether male and female, married or unmarried, might then be illuminating as a part of a larger story about the community of astronomers who form a particularly close-knit observatory family. Peggy A. Kidwell, "Cecilia Payne-Gaposchkin," in *Uneasy Careers and Intimate Lives*, ed. P. Abir-Am and D. Outram (Rutgers University Press, 1987), 226, 224. Likewise, the biography of the Australian radio astronomer Ruby Payne-Scott finishes with the strange accolade of "remarkable scientist and mother" after an entire book spent resurrecting Payne-Scott's scientific career. W. Miller Goss and R. X. McGee, *Under the Radar* (Springer, 2009), 263. If we accept the feminist critique that silence about scientist's familial life has led to, at best, an incomplete portrait of their working lives, then likewise relegating to a note that "Eddington displayed no interest in romantic relationships with women" surely similarly critiques many a great (unmarried) man's biography. Matthew Stanley, *Practical Mystic* (University of Chicago Press, 2007), 252.

65. K. Hufbauer, *Exploring the Sun* (Johns Hopkins University Press, 1991), 132–133. For Shapley's opposition, see David DeVorkin, *Science with a Vengeance* (Springer-Verlag, 1992), 234 n. 52; for Payne-Gaposchkin's, see Peggy Kidwell, "Harvard Astronomers and World War II," in *Science at Harvard University* (Associated University Presses, 1992), 288.

66. David DeVorkin, "Who Speaks for Astronomy?" *Historical Studies in the Physical and Biological Sciences* 31 (2000), no. 1, 52–92.

67. Bart Bok to Dr. Paul Buck, October 7, 1952, Harvard College Archives, UA V 630.28.5, p. 2.

68. Ibid. Harvard's 24-inch reflector was revived from an "almost worthless" status with the addition of a photoelectric photometer. In 1945, Harvard's three largest telescopes were not in operation. See "Report to H.S. on the status of Oak Ridge Station," September 25, 1945, Harvard College Archives, UA V 630.28.5, file Agassiz Station 1944–57.

69. Donald Menzel, "Memorandum to Observatory Personnel," March 1953, Harvard College Archives, UAV 630.28, box Bok, Administrative Files, 1946–July 1953, folder March–April 1953. Bart Bok to Dr. Paul Buck, October 7, 1952, Harvard College Archives, UA V 630.28.5, p. 2.

70. Greenstein, *Sources for History of Modern Astrophysics*, 31. At least in his memory, Greenstein is not clear on exactly whom he considered his advisor, mentioning both Menzel and Bok. See Jesse L. Greenstein, "An Astronomical Life," *Annual Review of Astronomy and Astrophysics* 22 (1984), 1–35, at 6.

71. B. J. Bok, "Some Notes About Finances and Personnel of the Harvard College Observatory," February 27, 1948, Harvard College Archives, UA V 630.28, B. J. Bok Administration Files, box 1944–July 53, file 1948.

72. Fred Whipple to Greenstein, August 11, 1949, Greenstein Papers, California Institute of Technology Archives, 43.8.

73. Donald Menzel, "The Scientific and Budgetary Future of Harvard College Observatory," February 15, 1952, copy in Harvard College Archives, UA V 630.28, B. J. Bok Administration Files, box 1944–July 53, file Jan–Feb 1952.

74. Harlow Shapley, "Draft of the Introduction to the Annual Report of the Observatory for 1951–52," ca. February 6, 1952, copy in Ira S. Bowen Papers, Huntington Library, box 25, folder 25.406.

75. Harlow Shapley's draft of February 1952 provided the Observatory Council with a picture of budgetary requirements. See copy in Harvard College Archives, UA V 630.28, B. J. Bok Administration Files, box 1944–July 53, file Jan–Feb 1952.

76. Minutes of the HCOC Meeting, April 11, 1952, Harvard College Archives, UA V 630.28, B. J. Bok Administration Files, box 1944–July 53, file March–May 1952.

77. Bart Bok to Dr. Paul Buck, October 7, 1952, Harvard College Archives, UA V 630.28.5, p. 4.

78. "Transcription of the remarks made at the meeting of the Visiting Committee of Harvard University," March 26, 1954, Harvard College Archives, UA V 630.28 box Bok, Administrative Files, August 1953–January 1957, folder Jan–March 1954, pages 3–4.

79. Unsigned letter to Mrs. Agassiz, April 3, 1952, Harvard College Archives, UA V 630.28, B. J. Bok Administration Files, box 1944–July 53, file March–May 1952.

80. *Report of the President of Harvard College and Reports of Departments,* vol. LI, no. 25, 1951–52, p. 315, Harvard College Archives.

81. Bart Bok to Harvard Observatory Council, October 19, 1954, Harvard College Archives, UA V 630.28.5.

82. B. J. Bok to Richard Perkin, July 16, 1954, Harvard College Archives, UA V 630.28, B. J. Bok Administration Files, box Aug 1953–1957, file April–July 1954.

83. Paul Forman, "Behind Quantum Electronics," *Historical Studies in the Physical and Biological Sciences* 18 (1987): 149–229.

84. Tentative draft of letter, January 1954, Harvard College Archives, UA V 630.28, B. J. Bok Administration Files, box Aug 1953–1957, file Aug-Dec 1953.

85. Bart Bok to Ryle, October 7, 1953, Ryle Papers, Churchill College, Cambridge University, box 1, folder 3.

86. Bart Bok to Donald Menzel, February 24, 1954, Harvard College Archives, UA V 630.28.5, file Agassiz Station 1944–57, p. 3.

87. Ronald Doel, *Solar System Astronomy in America* (Cambridge University Press, 1996), 196, 193.

88. H.C. van de Hulst, *A Course on Radio Astronomy* (Leiden, 1951).

89. Warwick, *Masters of Theory*, 24.

90. Bok to Jan Oort, October 6, 1953, Jan Oort Papers, Leiden University Library, box 151, folder d.

91. "Eighteenth Informal Memorandum from Harlow Shapley," August 8, 1952, Harvard College Archives UA V 630.28, B. J. Bok Administration Files, box 1944–July 53, file June–Sept 1952.

92. *Report of the President of Harvard College and Reports of Departments,* vol. LI, no. 26, 1952–53, p. 288, Harvard College Archives. This was also the title of Heeschen's dissertation.

93. Bart Bok to "Radio Astronomy Men," May 3, 1954, Harvard College Archives, UA V 630.28.5.

94. "Memorandum on Research Activities," B. J. Bok to the Visiting Committee of HCO, March 25, 1953, Harvard College Archives, UA V 630.28, B. J. Bok Administration Files, box 1944–July 53, file Mar–April 1953.

95. Bart Bok and Harold Ewen, "Radio Astronomy in the Microwave Region," November 10, 1953, Harvard College Archives, UA V 630.453.15, p. 5.

96. Bok to Greenstein, November 26, 1954, Greenstein Papers, California Institute of Technology, folder 3.26.

97. Bart J. Bok, "Radio Studies of Interstellar Hydrogen," *Sky and Telescope* 13 (October 1954): 408–412, 408.

98. B. J. Bok to Richard Perkin, July 16, 1954, Harvard College Archives, UA V 630.28, B. J. Bok Administration Files, box Aug 1953–1957, file April–July 1954.

99. Memo from Bok to Menzel, October 20, 1955, Harvard College Archives, UA V 630.28.5, file Agassiz Station 1944–57. As it turned out a generous gift was made to the Observatory for a "Mobile Unit" attached to the Agassiz Station to "contain an improved version of our present electronic apparatus." The advantage of the Unit was that Ewen would "be able to continue laboratory development . . . without having to interrupt astronomical research." Bart Bok to Dr. Charles Schauer, Research Corporation, March 30, 1954, Harvard College Archives, UA V 630.28.5, file Agassiz Station 1944–57.

100. Minutes of HCOC Meeting, December 12, 1956, Harvard College Archives, UA V 630.28, B. J. Bok Administration Files, box Aug 1953–1957, file Dec '56–Jan '57.

101. Robert H. Kargon, "Temple to Science," *Historical Studies in the Physical Sciences* 8 (1977): 3–31, at 4 and 10. Chemistry, John Servos argues, emphasized the adherence to disciplinary lines over the same time period. See Servos, *Physical Chemistry from Ostwald to Pauling* (Princeton University Press, 1990).

102. Donald Osterbrock, *Pauper and Prince* (University of Arizona Press, 1993), 221–224. See also Helen Wright, *Explorer of the Universe* (American Institute of Physics, 1994); Kevles, *The Physicists*, 156; Kargon, "Temple to Science."

103. Van de Hulst to Greenstein, April 29, 1950, Greenstein Papers, California Institute of Technology Archives, 41.1.

104. I. S. Bowen to Bush, April 10, 1952, I. S. Bowen Papers, Huntington Library, box 21, folder 21.341.

105. David Kaiser, "Cold War Requisitions, Scientific Manpower, and the Production of American Physicists after World War II," *Historical Studies in the Physical and Biological Sciences* 33 (2002), no. 1: 131–159.

106. Ibid.; McCray, *Giant Telescopes*, 36.

107. See David Kaiser, Making Theory," PhD thesis, Harvard University, 2000, 29.

108. Kaiser, "The Postwar Suburbanization of American Physics," 862.

109. Ibid., 871.

110. Robert Bacher to E. G. Bowen, June 13, 1952, DuBridge Papers, California Institute of Technology Archives, 35.2.

111. I. S. Bowen to Bush, August 21, 1953, I. S. Bowen Papers, Huntington Library, box 24, folder 24.392. We can compare Bowen's caution with the optimism of several cases of new technology in astronomy. Ronald Doel's disciplinary change explicitly caused by the rise of "specialized research instruments" and the "proliferation of patrons." Doel, *Solar System Astronomy in America* (Cambridge University Press, 1996), 88 and 191–193. According to David DeVorkin, astronomers themselves promoted the use of photoelectric cells. See DeVorkin, "Electronics in Astronomy," 1207 and 1217.

112. E. G. Bowen to Vannevar Bush, August 22, 1952, CRS C3830, A1/3/11/1, part 1.

113. Handwritten Notes. n.d. (ca. 1952/53), Greenstein Papers, California Institute of Technology Archives, 77.2. Note the distinct parallels between Greenstein's notes and those of the radio physicist Edward Bowen in Australia. While "radio astronomy has tended to grow up in radio laboratories which are not closely associated with astronomical observatories," Bowen wrote, "today there is no doubt that radio can make tremendous contributions to the study of astronomy. . . . The time appears ripe, therefore, to bring the radio and visual observations into closer contact." E. G. Bowen, "Draft Programme for a Radio Observatory," May 1952, CRS C3830, A1/3/11/1.

114. Bowen to Bush, November 13, 1945, I. S. Bowen Papers, Huntington Library, box 1, folder 1.6.

115. Handwritten notes, n.d. (ca. 1952/53), Greenstein Papers, California Institute of Technology Archives, 77.2.

116. DuBridge to E. G. Bowen, February 21, 1952, DuBridge Papers, California Institute of Technology Archives, 35.2.

117. Bush to DuBridge, February 14, 1952, DuBridge Papers, California Institute of Technology Archives, 35.2.

118. Robert Bacher to E. G. Bowen, April 1, 1952, copy in I. S. Bowen Papers, Huntington Library, box 22, folder 22.346.

119. Bush to I. S. Bowen, March 17, 1952, I. S. Bowen Papers, Huntington Library, box 21, folder 21.341.

120. I. S. Bowen to Bush, March 11, 1952, I. S. Bowen Papers, Huntington Library, box 21, folder 21.341.

121. I. S. Bowen to Tuve, March 24, 1952, I. S. Bowen Papers, Huntington Library, box 23, folder 23.383.

122. E. G. Bowen to DuBridge, June 20, 1952, DuBridge Papers, California Institute of Technology Archives, 108.8.

123. Handwritten notes, ca. 1952/53) Greenstein Papers, California Institute of Technology Archives, 77.2.

124. Bowen to Bush, September 25, 1952, I. S. Bowen Papers, Huntington Library, box 21, folder 21.341.

125. Bacher to Greenstein, June 16, 1953, Greenstein Papers, California Institute of Technology Archives, 77.2.

126. Bowen to Bush, October 19, 1955, I. S. Bowen Papers, Huntington Library, box 30, folder 30.491.

127. Memo from Greenstein to Bacher, "Proposed revision of Graduate Curriculum in Astronomy," December 15, 1955, I. S. Bowen Papers, Huntington Library, folder 30.494.

128. DuBridge to E. R. Piore, September 21, 1954, DuBridge Papers, California Institute of Technology Archives, 35.2. DuBridge to Elliot Montroll, September 17, 1955, DuBridge Papers, California Institute of Technology Archives, folder 35.2.

129. DuBridge to Bolton, October 8, 1954, DuBridge Papers, California Institute of Technology Archives, folder 35.2.

130. Draft of Proposal to the Office of Naval Research for a Project at the California Institute of Technology in Radio Astronomy, December 14, 1954, I. S. Bowen Papers, Huntington Library, box 28, folder 28.467.

131. DuBridge to E. R. Piore, September 21, 1954, DuBridge Papers, California Institute of Technology Archives, folder 35.2. For further biographical details, see K. I. Kellermann, "John Gatenby Bolton (1922–1993)," *Publications of the Astronomical Society of the Pacific* 108 (1996): 729–737. For a hint at how significant Bolton's initial identifications were for astronomy, see Otto Struve, "Progress in Radio Astronomy—II," *Sky and Telescope* 9 (1950): 55–56.

132. Bolton to Greenstein, 12–11–54, Greenstein Papers, California Institute of Technology Archives, 77.2.

133. Greenstein to van de Hulst, December 30, 1954, Greenstein Papers, California Institute of Technology Archives, 41.1.

134. Bacher to Weaver, July 22, 1955, DuBridge Papers, California Institute of Technology Archives, 35.3.

135. McCray, *Giant Telescopes*, 19.

136. Handwritten notes, n.d. (ca. 1952/53), Greenstein Papers, California Institute of Technology Archives, 77.2.

137. Handwritten notes, n.d. (ca. 1952/53), Greenstein Papers, California Institute of Technology Archives, 77.2.

138. News Release for December 12, 1952, folder 37, box 7, Record Group 205D. RAC. "If it can be shown that the radio and optical diameters agree and that the position of the astronomical object coincides with that of the radio source within the error of measurement of

that source . . . positive identification [is] possible." From 'Radio Astronomy," *Engineering and Science* 16 (January 1953): 32, 36.

139. T. K. Menon to Jan Oort, August 12, 1956, Jan Oort Papers, Leiden University Library, box 109, folder d.

140. Greenstein to Bolton, April 30, 1956, Greenstein Papers, California Institute of Technology Archives, 77.2.

141. "A. J. Higgs' talk on Australian radio" (transcript accompanying letter from Higgs to J. Pratt (Australian Broadcasting Corporation, Talks Department), August 11, 1954, CRS C3830, D10/2, part 1.

CHAPTER 4

1. Ian Wark, "Physical Sciences," in *Science in Australia* (Cheshire, 1952), 38.

2. Harlow Shapley, "Report for IAU Commission 39," January 14, 1955, O. Struve Papers, 81/35, Bancroft Library, UCB, box 3, folder IAU-General +1955.

3. Warren Weaver Correspondence, August 24, 1955, folder 86, box 7, Series 410D, Record Group 1.2, Rockefeller Foundation Archives, RAC.

4. Agar, *Science and Spectacle*, 55.

5. "A Proposal for a 250-ft Aperture Steerable Paraboloid for Use in Radio Astronomy" (n.d.), Ryle Papers, Churchill College, Cambridge University, box 7-15.

6. "Minutes of the 1st Meeting of the 'Committee for Radio-Astronomy,' February 10, 1950, Ryle Papers, Churchill College, Cambridge University, box 7-15.

7. Lovell, *The Story of Jodrell Bank*, 29.

8. Ibid., 33–34.

9. Ryle to Sir Lawrence Bragg, March 23, 1955, Ryle Papers, Churchill College, Cambridge University, box 1-3.

10. Lovell, *Astronomer by Chance*, 216–217.

11. Agar, *Science and Spectacle*, 68.

12. Lovell, *The Story of Jodrell Bank*, 41.

13. Bowen to White, June 6, 1952, CRS C3830, A1/3/11/3, part 1.

14. Lovell, *The Story of Jodrell Bank*, 83.

15. Minutes of Planning Committee—Giant Radio Telescope, November 6, 1954, CRS C3830, A1/3/11/2.

16. Agar, *Science and Spectacle*, 195–196.

17. Lovell, *Astronomer by Chance*, 284–286.

18. A. C. B. Lovell, "Radio Astronomy at Jodrell Bank—I," *Sky and Telescope* 12 (1953), no. 136: 94–96, 114.

19. Bowen to DuBridge, August 30, 1952, CRS C3830, A1/3/11/1, part 1.

20. Pawsey to Mark Oliphant, September 10, 1951, CRS C3830, A1/1/4. Oliphant certainly supported the Radiophysics Lab's drive to preserve its place, going so far as to argue that Stromlo should expand, not into radio astronomy, but into solar astronomy in order to support the Radiophysics Lab's work on solar noise. See M. L. Oliphant, "Solar Physics in Australia" (n.d.), CRS, C3830, A1/1/4.

21. Bowen to Oliphant, July 3, 1952, CRS C3830, A1/3/11/3, part 1.

22. Minutes of Radio Astronomy Committee Meeting, December 12, 1952, CRS C3830, B2/2, part 2.

23. White to Sir Edward Appleton, June 27, 1952, CRS C3830, A1/3/1/3, part 1. Prestige, attached to scientific leadership within a field, continued to be a major selling point to the Australian government after radio astronomy was institutionally secure within the Division of Radiophysics. "A major part in the development of radio astronomy has taken place here in Australia and it can safely be said that this country is in the forefront in this particular science." Bowen to White, April 27, 1954, CSIRO Series 9, G23/2/[1/15]. "The X General Assembly of the International Scientific Radio Union," *Australian Journal of Science* 31 (1952), no. 3: 91–92, at 92.

24. "Meeting: Giant Radio Telescope," November 10, 1952, CRS C3830, A1/3/11/2.

25. J. L. Pawsey, "Notes on the Cosmic Noise Programme in Radiophysics," February 19, 1951, CRS C3830, A1/1/1, part 6.

26. "Large Radio Telescope," September 23, 1952. AA C3830, A1/3/11/2; December 4, 1952, CRS C3830, A1/3/11/2.

27. Minutes of Radio Astronomy Committee Meeting, December 12, 1952, CRS C3830, B2/2, part 2.

28. White to Tizard, October 6, 1953, CSIRO Series 9, G23/2/[1/15].

29. Bowen to White, March 26, 1953, CRS C3830, E2/2, part 2.

30. Pawsey, "Radio Astronomy Group—Current Programme," July 8, 1954, CRS C3830, A1/1/1, part 9.

31. Ian Wark, "Physical Sciences," in *Science in Australia* (Cheshire, 1952), 38.

32. White to Bowen, July 17, 1952, CRS C3830, A1/3/11/1.

33. Bowen to White, June 17, 1952, CRS C3830, A1/3/11/3, part 1.

34. Agar, *Science and Spectacle*, 60.

35. Bush to Bowen, September 3, 1952, CRS C3830, A1/3/11/1, part 1.

36. Bowen to Vannevar Bush, August 22, 1952, CRS C3830, A1/3/11/1, part 1.

37. Bowen to R. G. Casey, February 6, 1954, CSIRO Series 3, folder PH/BOW/9B, part 2.

38. Ibid.

39. Bowen to Guy Gresford, Secretary, CSIRO, October 27, 1953, CSIRO Series 3, PH/BOW/9B.

40. "Statement by the President and Secretary of the International Astronomical Union in support of construction of a giant radio telescope in Australia," July 28, 1954, CRS C3830, A1/3/11/7. Five days earlier, a "Statement by the Members of Council of the Royal Astronomical Society support of construction of a giant radio telescope in Australia" appeared; one assumes a connection, since the style, format, and timing of the two are so coincident. The RAS was firm in its belief that the crucial benefit was the completion of coverage in the southern hemisphere by an instrument similar in ability to the Manchester one. July 23, 1954, CRS C3830, A1/3/11/7.

41. Tuve to Bowen, May 4, 1954. AA C3830, A1/3/11/1, part 1; Bush to Bowen, June 9, 1954, CRS C3830, A1/3/11/1, part 1; Robertson, *Beyond Southern Skies*, 118.

42. White to Tizard, April 8, 1954, CRS C3830, A1/3/11/1, part 1.

43. M. L. Oliphant to Bowen, June 12, 1952, CRS C3830, A1/3/11/3, part 1.

44. Bush to White, October 13, 1954, CRS C3830, A1/3/11/3, part 2.

45. White to Bush, October 13, 1954, CRS C3830, A1/3/11/3, part 2.

46. Nicholls to Pawsey, May 23, 1955, CRS C3830, A1/3/11/3, part 3.

47. Bowen to Tuve, March 16, 1955, CRS C3830, A1/3/11/3, part 3.

48. Casey to Bowen, August 2, 1955, CRS C3830, A1/3/11/3, part 3.

49. R. G. Casey to G. J. Coles, June 14, 1955, folder 86, box 7, Series 410D, Record Group 1.2, Rockefeller Foundation Archives, RAC.

50. Bowen to Tuve, March 16, 1955, CRS C3830, A1/3/11/3, part 3.

51. Tuve to Bowen, March 3, 1955, CRS C3830, A1/3/11/3, part 3.

52. Bowen to Clunies-Ross, March 10, 1955, CRS C3830, A1/3/11/3, part 3.

53. W.W. (Warren Weaver) Correspondence, August 24, 1955, folder 86, box 7, Series 410D, Record Group 1.2, Rockefeller Foundation Archives, RAC.

54. Robertson, *Beyond Southern Skies*, 154–186.

55. Ryle to Lovell, August 7, 1950, Ryle Papers, Churchill College, Cambridge University, box 1-1.

56. "Routine Observations in Radio Astronomy at Cambridge," February 18, 1950, Ryle Papers, Churchill College, Cambridge University, box 1-1.

57. Edge and Mulkay, *Astronomy Transformed*, chapter 5.

58. Ibid., 142.

59. M. Ryle, "The Mullard Radio Astronomy Observatory, Cambridge," *Nature* 180 (July 1957): 110–112, 110.

60. M. Ryle and J. A. Ratcliffe, "Radio Astronomy," *Endeavor* XI (1952): 117–125, 118. The illustration of Jodrell Bank's resolution making the sun 'blurred' was earlier deployed by J. A. Ratcliffe at one of the Royal Institution of Great Britain Weekly Evening Meetings, Friday, November 23, 1951. His talk was titled 'Radio Astronomy.' Ryle repeated his basic criticisms against Lovell and Jodrell bank—though rarely mentioning names—several times in the early 1950s. See M. Ryle, "Radio Astronomy," *Reports on Progress in Physics* 13 (1950): 184–246.

61. Ryle and Ratcliffe, "Radio Astronomy," *Endeavor* 11 (1952), 119.

62. A. C. B. Lovell, "Radio Astronomy at Jodrell Bank—I," *Sky and Telescope* 12 (1953), no. 136: 94–96, 114.

63. "Notes on the projected 200–250ft Aperture Paraboloid" (n.d.), Ryle Papers, Churchill College, Cambridge University, box 1-1.

64. "A Proposal for a 250-ft Aperture Steerable Paraboloid for Use in Radio Astronomy" (n.d.), Ryle Papers, Churchill College, Cambridge University, box 7-15.

65. M. Ryle, "The Requirements of Radio Astronomy at Cambridge," April 1954, Ryle Papers, Churchill College, Cambridge University, box 7-14.

66. Ryle, "The Mullard Radio Astronomy Observatory, Cambridge," 110.

67. Ibid., 112.

68. Speech of Mr. Mullard at the Opening of the Mullard Radio Astronomy Observatory, Cambridge, 1957, Ryle Papers, Churchill College, Cambridge University, box 7-14.

69. "The radio telescope has therefore shown itself to be an important adjunct to the world's greatest optical telescope," said Sir Edward Appleton at the end of his 1954 Presidential Address to the British Association for the Advancement of Science. Sir Edward Appleton, "Science for Its Own Sake," *Science* 119 (January 22, 1954): 103–109, 108.

70. Ryle to T. E. Goldup, Chairman of the Board, Mullard Corporation, June 9, 1955, Ryle Papers, Churchill College, Cambridge University, box 1-3.

71. Edge and Mulkay, *Astronomy Transformed*, 24.

72. Ryle to Oort, March 23, 1956, Oort Papers, University of Leiden library, box 28, folder j.

73. Oort to Pawsey, March 27, 1956, Oort Papers, University of Leiden library, box 109, folder d.

74. Lovell to Oort, February 22, 1956, Oort Papers, University of Leiden library, box 109, folder c.

75. Ryle, "Radio Astronomy: The Cambridge Contribution," in *Search and Research*, 23.

76. Oral history interview, Bernard Mills, Australian National Library, DeB 584, Transcript p. 7; B. Y. Mills and A. Little, "A High-Resolution Aerial System of a New Type," *Australian Journal of Physics* 6 (1953): 272–278.

77. W. N. Christiansen and B. Y. Mills, "Dr. J. L. Pawsey," April 1964, Pawsey Papers, Basser Library, Australian Academy of Science Archives, MS20.

78. W. N. Christiansen, "History of Australian Astronomy: History and Propaganda in Astronomy," *Proceedings of the Astronomical Society of Australia* 8 (1989), no. 1: 96–101.

79. "Minutes of Radio Telescope Planning Meeting," June 29, 1955, CRS C3830, A1/3/11/2.

80. Bowen to White, August 20, 1956, CRS C3830, E2/2. part 2.

81. Pawsey, "Radio Astronomy Group—Current Programme," July 8, 1954, CRS C3830, A1/1/1, part 9.

82. Pawsey to Oort, March 14, 1955, Jan Oort Papers, University of Leiden Library, box 109, folder d.

83. Greenstein to Bowen, May 27, 1954, Greenstein Papers, California Institute of Technology Archives, folder 4.8. DuBridge to E. G. Bowen, June 9, 1954, CRS CRS3830, A1/3/11/1, part 1. "At the present time we do not see that funds in sight which will be required to construct a very large aerial of the steerable searchlight type though a small dish for 21 centimeter work might well be within our program." DuBridge to Bolton, October 8, 1954, DuBridge Papers, California Institute of Technology Archives, 35.2.

84. Greenstein to Bowen, May 27, 1954, Greenstein Papers, California Institute of Technology Archives, 4.8.

85. DuBridge to Bolton, December 30, 1954, DuBridge Papers, California Institute of Technology Archives, 35.2.

86. Greenstein, *Sources for History of Modern Astrophysics*, 157–158; Harvey M. Sapolsky, *Science and the Navy* (Princeton University Press, 1990).

87. Bacher to Weaver, July 22, 1955, DuBridge Papers, California Institute of Technology Archives, 35.3.

88. DuBridge to Tuve, March 7, 1955, DuBridge Papers, California Institute of Technology Archives, 35.3.

89. Bolton to Greenstein, ca. Feb 1955, Greenstein Papers, California Institute of Technology Archives, 77.2.

90. Release from the News Bureau, California Institute of Technology, April 3, 1956, Greenstein Papers, California Institute of Technology Archives, 77.2.

91. "A proposal to the Office of Naval Research and the Rockefeller Foundation for a radio astronomy facility at the California Institute of Technology," July 20, 1955, folder 86, box 7, Series 410D, Record Group 1.2, Rockefeller Foundation Archives, RAC, p. 6.

92. Warren Weaver Correspondence, August 24, 1955, folder 86, box 8, folder 86, box 7, Series 410D, Record Group 1.2, Rockefeller Foundation Archives, RAC.

93. "A Proposal to the Office of Naval Research and the Rockefeller Foundation for a radio astronomy facility at the California Institute of Technology," July 20, 1955, folder 86, box 7, Series 410D, Record Group 1.2, Rockefeller Foundation Archives, RAC, p. 6. The Caltech radio telescope's "versatility" and "uniqueness" were also proclaimed to the media. See news release by CIT News Bureau, September 25, 1958, Greenstein Papers, California Institute of Technology Archives, 77.2.

94. "A proposal to the Office of Naval Research and the Rockefeller Foundation for a radio astronomy facility at the California Institute of Technology," July 20, 1955, folder 86, box 7, Series 410D, Record Group 1.2, Rockefeller Foundation Archives, RAC, pp. 3–4.

95. Bolton to Greenstein, December 30, 1954, Greenstein Papers, California Institute of Technology Archives, 77.2.

96. "A Proposal to the Office of Naval Research and the Rockefeller Foundation for a radio astronomy facility at the California Institute of Technology," July 20, 1955, folder 86, box 7, Series 410D, Record Group 1.2, Rockefeller Foundation Archives, RAC, p. 2. Bernard Mills' cross antenna in Australia, operating at a specific frequency, cost about US$80,000; the 250-foot steerable Manchester paraboloid would eventually consume around US$2 million.

97. Tuve to DuBridge, March 10, 1955, DuBridge Papers, California Institute of Technology Archives, 35.3.

98. Draft of Proposal to the Office of Naval Research for a Project at the California Institute of Technology in Radio Astronomy, December 14, 1954, Huntington Library, I. S. Bowen Papers, box 28, folder 28.467.

99. Greenstein, *Sources for History of Modern Astrophysics*, 115.

100. Ibid., 118,.

101. Circular letter from Struve, March 25, 1954, Greenstein Papers, California Institute of Technology Archives, 36.18.

102. Greenstein to Struve, May 17, 1954, Greenstein Papers, California Institute of Technology Archives, 36.18.

103. E. G. Bowen to DuBridge, January 15, 1958, DuBridge Papers, California Institute of Technology Archives, 108.8.

104. Bolton to Greenstein, April 23, 1956, Greenstein Papers, California Institute of Technology Archives, 77.2.

105. Greenstein to Bolton, April 30, 1956, Greenstein Papers, California Institute of Technology Archives, 77.2.

106. Release from News Bureau, California Institute of Technology, April 3, 1956, Greenstein Papers, California Institute of Technology Archives, 77.2.

107. I. S. Bowen to DuBridge, March 23, 1956, DuBridge Papers, California Institute of Technology Archives, folder 23.8. Ira Bowen's solution was to hire another stewardess and build another cottage for her.

108. Greenstein, *Sources for the History of Modern Astrophysics*, 125.

109. Greenstein to Bacher, September 3, 1958, Greenstein Papers, California Institute of Technology Archives, 2.11.

110. K. C. Westfold, *A Course in Radio Astronomy* (California Institute of Technology, 1959).

111. Greenstein to Tuve, June 5, 1957, Greenstein Papers, California Institute of Technology Archives, 39.12.

112. Kellermann, "John Gatenby Bolton," 735–736.

113. Greenstein to Maarten Schmidt, February 19, 1959, Greenstein Papers, California Institute of Technology Archives, 34.3.

114. Hanbury Brown, *Boffin*, 109–110.

115. Lovell to Manchester Vice-Chancellor, May 3, 1957, John Rylands Library/Jodrell Bank Archives, ACC—box 55, file 2.

116. Transcript of Interview with Bernard Lovell, July 6, 1971, John Rylands Library/Jodrell Bank Archives, Astronomy Transformed—box 3, file 2.

117. Lovell to Miles Clifford, June 27, 1957, John Rylands Library/Jodrell Bank Archives, ACC—box 55, file 2.

118. Lovell, *The Story of Jodrell Bank*, 160.

119. Ibid., chapter 35.

CHAPTER 5

1. Jesse Greenstein, oral history interview by Spencer Weart, April 7, 1977, Sources for History of Modern Astrophysics, 128.

2. Tuve to Bok, June 29, 1956, Greenstein Papers, California Institute of Technology Archives, 111.5.

3. Patrick McCray, *Giant Telescopes* (Harvard University Press, 2006), 291.

4. Allan Needell, "Nuclear Reactors and the Founding of Brookhaven National Laboratory," *Historical Studies in the Physical Sciences* 14 (1984): 93–122; Robert Crease, *Making Physics* (University of Chicago Press, 1999).

5. Lloyd Berkner to Raymond Seeger, November 8, 1954, Tuve Papers, Library of Congress, box 326. This realization was first made by Ira Bowen and Vannevar Bush during the planning for the new Palomar Observatory in 1946. They outlined the appointment of visitors to Palomar to maximize the telescope's use. In fact, the combination of Mount Wilson with Palomar had essentially doubled the instrument capacity of the observatory with no appreciable increase in staff, and students from Caltech found themselves having almost exclusive rights to the 60-inch Schmidt, and even regularly the 100-inch Hooker. Palomar started receiving visitors after 1949, establishing a sharing arrangement that remains a centerpiece of every new telescope.

6. Quoted from Jon Agar, *Science and Spectacle* (Harwood, 1998), 55.

7. John Krige, "The Installation of High-Energy Accelerators in Britain after the War: Big Equipment but not Big Science," in *The Restructuring of Physical Science in Europe and the United States, 1945–60*, ed. M. De Maria (World Scientific, 1989), 498. As early as 1946, a visiting French biologist lamented the passing of a culture of small, personal science. He "used to think that my work is something rare, highly personal[,] valuable," but on a trip through the United States he witnessed "350 biologists" working in "very big laboratories, huge library, three seminars a week, impressive organization, etc." Jean-Paul Gaudillière, "Paris–New York Roundtrip," *Studies in the History and Philosophy of Biological and Biomedical Sciences* 33 (2002), no. 3: 389–417, at 390.

8. Daniel Greenberg easily lumped the radio astronomers' new giant radio telescopes into a category alongside "oceanographic vessels [and] nuclear accelerators." *The Politics of American Science* (Penguin, 1969), 36.

9. B. Hevly, "Big Science and Big History," in *Big Science*, ed. P. Galison and B. Hevly (Stanford University Press, 1992), 356–357. Alvin Weinberg, "Impact of Large-Scale Science on the United States," *Science* 134:3473 (1961): 161–64. Daniel Kevles, "Big Science and Big Politics in the United States," *Historical Studies in the Physical and Biological Sciences* 27:2 (1997): 269–298. James Capshew and Karen Rader, "Big Science," *Osiris*, second series, 7 (1992): 3–25.

10. Alvin Weinberg, "Impact of Large-Scale Science on the United States," *Science* 134 (1961), no. 3473: 161–164.

11. Leila Bram, Acting Head Mathematics Branch, ONR, to Greenstein, November 23, 1953, Greenstein Papers, California Institute of Technology Archives, 27.2.

12. "Announcement of the Office of Naval Research. Contracts for Research in Astronomy and Astrophysics," December 7, 1954, Greenstein Papers, California Institute of Technology Archives, 27.2.

13. "Minutes of Meeting of Advisory Panel on Radio Astronomy, NSF," November 20, 1954, Tuve Papers, Library of Congress, 329, "Interim papers, Advisory Panel, NSF, 1954–55."

14. Proposal to NSF to sponsor a conference on Radio Astronomy, Draft, ca. March 19, 1953, Greenstein Papers, California Institute of Technology Archives, 113.1.

15. Greenstein to Bok, April 9, 1953, DuBridge Papers, California Institute of Technology Archives, 34.15.

16. Greenstein, oral history interview, 127.

17. Copy of letter from Gerard P. Kuiper, Yerkes Observatory, to Raymond Seeger, NSF, March 20, 1953, Greenstein Papers, California Institute of Technology, 112.1.

18. Keller and Keenan to Greenstein, March 31, 1953, Greenstein Papers, California Institute of Technology Archives, 112.1.

19. Greenstein to Tuve, November 16, 1954, Bart Bok Papers, Harvard College Archives, box Aug 1953–1957, file Nov–Dec 1954.

20. "Appendix B: On the Need for a Radio Astronomy Program in the United States," Richard Emberson, AUI, to Raymond Seeger, NSF, April 21, 1955, Tuve Papers, Library of Congress, 325.

21. Donald Menzel, "Survey of the Potentialities of Cooperative Research in Radio Astronomy," April 13, 1954, Tuve Papers, Library of Congress, 329, "Interim Papers, Advisory Panel, NSF 1954–55."

22. Greenstein to Hagen, June 1, 1954, Greenstein Papers, California Institute of Technology Archives, 14.10.

23. Peter van de Kamp, "Report to the National Science Foundation of the Conference on Astronomical Research in Small Academic Departments held at Swarthmore College," April 19–21, 1954, Greenstein Papers, California Institute of Technology Archives, 112.2, p. 11.

24. Struve to Greenstein, April 29, 1955, Greenstein Papers, California Institute of Technology Archives, 36.18.

25. Harvard College Observatory Council minutes, April 11, 1954, Bart Bok Papers, Harvard College Archives, box Aug 1953–Jan 57, file April–July 1954.

26. Greenstein to Whitford, May 17, 1954, Greenstein Papers, California Institute of Technology Archives, 43.9.

27. Correspondence from Jesse Greenstein dated May 14, 1954, "Minutes of the Radio Astronomy Conference, May 20, 1954 at the Empire State Building," Tuve Papers, Library of Congress, 326.

28. Berkner to Seeger, November 8, 1954, Tuve Papers, Library of Congress, 326.

29. Bart Bok, "Evaluation of the NSF Research Proposal—Feasibility Study of a National Radio Astronomy Facility," August 30, 1954, Tuve Papers, Library of Congress, 329, "AUI & Radio Ast., Spring 1955."

30. "Minutes of the Radio Astronomy Conference, May 20, 1954 at the Empire State Building," Tuve Papers, Library of Congress, 326.

31. Struve to Greenstein, April 29, 1955, Greenstein Papers, California Institute of Technology Archives, 36.18.

32. "Minutes of the Conference on Radio Astronomy, July 11, 1956," Leo Goldberg Papers, Harvard College Archives, box "L.G. Corres. with NRAO," file "Steering Committee—AUI Radioastronomy Facility."

33. Greenstein to Hagen, June 1, 1954, Greenstein Papers, California Institute of Technology Archives, 14.10.

34. M. B. Karelitz, "The 250-foot diameter reflector being built for the University of Manchester, England," March 25, 1955, Tuve Papers, Library of Congress, 326.

35. Bowen to Tuve, June 28, 1956, Greenstein Papers, California Institute of Technology Archives, 111.5.

36. Tuve to DuBridge, March 10, 1955, Tuve Papers, Library of Congress, 329, "AUI & Radio Ast, Spring 1955."

37. Talk by Lloyd Berkner in "Minutes of the Conference on Radio Astronomy, 11 July 1956," I. S. Bowen Papers, Huntington Library, 35.565, on 37–38.

38. Tuve to I. S. Bowen, March 14, 1952, I. S. Bowen Papers, Huntington Library, 23.383.

39. Alan Needell, "Lloyd Berkner, Merle Tuve, and the Federal Role in Radio Astronomy," *Osiris* 3 (1987): 261–288.

40. Allan A. Needell, *Science, Cold War and the American State* (Harwood, 2000), 278 and chapters 3–5.

41. Alan A. Needell, "The Carnegie Institution of Washington and Radio Astronomy: Prelude to an American National Observatory," *Journal for the History of Astronomy* 22 (1991): 55–67, at 62.

42. Ibid., at 56.

43. Bart Bok, "Research Objectives for Large Steerable Paraboloid Radio Reflectors," April 22, 1955, Tuve Papers, Library of Congress, 329, folder "AUI & Radio Ast., Spring 1955."

44. Ibid.

45. Memorandum from J. L. Greenstein and R. Minkowski to the Members of the NSF Committee on Radio Astronomy. May 18, 1955, copy in Tuve Papers, Library of Congress, box 327.

46. Bart Bok, "Research Objectives for Large Steerable Paraboloid Radio Reflectors," April 22, 1955, Tuve Papers, Library of Congress, 329, folder "AUI & Radio Ast., Spring 1955."

47. Greenstein to Bok, April 8, 1955, Greenstein Papers, California Institute of Technology Archives, 3.26.

48. Memorandum from Merle Tuve, acting for the Carnegie radio astronomy group, to Bart Bok, April 14, 1955, Tuve Papers, Library of Congress, 329, "AUI & Radio Ast., Spring 1955."

49. Greenstein to Menzel, July 27, 1956, Greenstein Papers, California Institute of Technology Archives, 23.4.

50. "Draft. Planning Document for the Establishment and Operation of a National Radio Astronomy Facility, IV:2," May 1956, Tuve Papers, Library of Congress, 324.

51. Purcell to Tuve, June 27, 1956, Leo Goldberg Papers, Harvard College Archives, box "L.G. Corres. with NRAO," file "Steering Committee—AUI Radioastronomy Facility."

52. Hagen to Chairman and Members of the HSF Panel on Radio Astronomy, May 10, 1955, Greenstein Papers, California Institute of Technology, 111.6.

53. Ibid.

54. Greenstein to Bok, April 8, 1955, Greenstein Papers, California Institute of Technology Archives, 3.26.

55. Hagen to Chairman and Members of the HSF Panel on Radio Astronomy, May 10, 1955, Greenstein Papers, California Institute of Technology Archives, 111.6.

56. Goldberg to Bok, April 14, 1955, Leo Goldberg Papers, Harvard College Archives, box "L.G. Corres. with NRAO," file "Steering Committee—AUI Radioastronomy Facility."

57. Handwritten Notes, ca. end 1955/ beg 1956, Greenstein Papers, California Institute of Technology Archives, 111.6.

58. Quote from Richard Emberson. Needell, *Science, Cold War and the American State*, 284. See also James Bamford, *The Puzzle Palace* (Penguin, 1983).

59. Greenstein and Minkowski to the Members of the NSF Committee on Radio Astronomy, May 18, 1955, Tuve Papers, Library of Congress, 327.

60. Greenstein and Minkowski to Members of NSF Committee on Radio Astronomy, May 18, 1955, Greenstein Papers, California Institute of Technology Archives, 111.6.

61. Ibid.

62. Bok to Tuve, cc. Hagen, Greenstein, Kraus, Minkowski, Purcell, April 28, 1955, Greenstein Papers, California Institute of Technology Archives, 111.6.

63. "Draft. Planning Document for the Establishment and Operation of a National Radio Astronomy Facility, I:3–4," May 1956, Tuve Papers, Library of Congress, box 324. "We cannot investigate many of the more urgent problems in radio astronomy because we do not have the necessary equipment," said the document.

64. "Draft. Planning Document for the Establishment and Operation of a National Radio Astronomy Facility, I:6–7," May 1956, Tuve Papers, Library of Congress, box 324.

65. Seeger to Tuve, "Excerpts from the NSF Budget," November 10, 1955, Tuve Papers, Library of Congress, 327, "AUI-NSF Sept 1955→."

66. Bok to Tuve, July 3, 1956, Greenstein Papers, California Institute of Technology Archives, 111.5.

67. Comments by Donald Menzel, "Minutes of the Conference on Radio Astronomy," July 11, 1956, Ira S. Bowen Papers, Huntington Library, 35.565, 43.

68. Greenstein to Tuve, July 6, 1956, Greenstein Papers, California Institute of Technology Archives, 111.6.

69. Menzel to Bowen, October 3, 1956, Ira S. Bowen Papers, Huntington Library, 34.561.

70. Bowen to Menzel, October 10, 1956, Ira S. Bowen Papers, Huntington Library, 34.561.

71. Hagan to Waterman, March 23, 1956, Tuve Papers, Library of Congress, 327, "NSF Panel—AUI Spring '56."

72. "Memorandum from Menzel to the Members of the Committee to Nominate a Director for the National Radio Observatory: I. S. Bowen, W. W. Morgan, C. D. Shane, O. Struve, and J. B. Wiesner," October 18, 1956, Ira S. Bowen Papers, Huntington Library, 34.561.

73. Struve to Menzel, October 8, 1956, Ira S. Bowen Papers, Huntington Library, 34.561; Wiesner to Menzel, October 17, 1956, Ira S. Bowen Papers, Huntington Library, 34.561.

74. Goldberg to Berkner, November 30, 1956, Ira S. Bowen Papers, Huntington Library, 34.549.

75. Struve to Whitford, January 7, 1961, Leo Goldberg Papers, Harvard College Archives, 8, "AUI Gen-Corres."

76. Struve to Rabi, May 12, 1961, Leo Goldberg Papers, Harvard College Archives, box "L.G. Corres. with NRAO," file "NRAO."

77. Berkner to Dr. Augustus B. Kinzel, Union Carbide Corp, October 17, 1960, Leo Goldberg Papers, Harvard College Archives, box "L.G. Corres. with NRAO," file "NRAO 140' and 300' Telescopes." A. Weinberg, "Impact of Large-Scale Science on the United States," *Science* 134 (1961), no. 3473: 161–164.

78. Struve to Rabi, October 31, 1961, Otto Struve Papers, Bancroft Library, University of California, Berkeley, 67/133, 1, "Outgoing Letters, 1949–61."

79. Goldberg to Rabi, October 18, 1961, Leo Goldberg Papers, Harvard College Archives, 8, "AUI Gen-Corres."

CONCLUSION

1. *The Dish*, directed by Rob Sitch (Working Dog Pictures, 2000).

2. Hanbury Brown, *Boffin*, 139.

3. Sullivan, *Cosmic Noise*, 429.

4. Frank Edmonson, *AURA and Its US National Observatories* (Cambridge University Press, 1997); Joseph Tatarewicz, *Space Technology and Planetary Astronomy* (Indiana University Press, 1990); Ronald E. Doel, *Solar System Astronomy in America* (Cambridge University Press, 1996).

5. Pamela O. Long, *Openness, Secrecy, Authorship* (Johns Hopkins University Press, 2004); B. J. Hunt, "Michael Faraday, Cable Telegraphy and the Rise of Field Theory," *History of Technology* 13 (1991): 1–19; Peter Galison, *Einstein's Clocks, Poincaré's Maps* (Norton, 2004).

6. Sagan once claimed that the definition of "an advanced civilization" was "one able to engage in long-distance radio communication using large radio telescopes." Sagan, *The Cosmic Connection*, 221.

7. Lee A. DuBridge, "The Effects of World War II on the Science of Physics," *American Journal of Physics* 17 (1949): 273–281, at 273–275.

8. Matthew H. Wisnioski, *Engineers for Change* (MIT Press, 2012).

9. I am thinking of particularly John Le Carré's *The Spy Who Came In From The Cold* (Scribner, 2001), Neil Sheehan's *A Bright Shining Lie* (Vintage, 1989), Stephen E. Ambrose and Douglas Brinkley's *Rise to Globalism* (Penguin, 1985), and Elaine Tyler May's *Homeward Bound* (Basic Books, 1999).

10. In the foreword to the 2000 edition of *The Cosmic Connection*, Carl Sagan's wife, Ann Druyan, expressed fear that activist scientists would distort science for political imperatives, citing "Lysenko, Mengele, and Teller" as easy exemplars. To put Edward Teller (the leader of the American hydrogen bomb project) in the same league as the fraudulent Soviet biologist and the Nazi human experimenter might appear unjustifiable. Druyan claimed that the rubble of Nagasaki tarnished Sagan's image of science, and that the political ambitions of Teller and the hydrogen bomb require an ethical response from scientists and the wider public. Ann Druyan, "A New Sense of the Sacred," in Carl Sagan, *Carl Sagan's Cosmic Connection*, ed. J. Agel (Cambridge University Press, 2000), xxii–xxiii.

11. "Almost half of the scientists on the planet, it had been estimated, were employed by one or another of the almost two hundred military establishments worldwide. And they were not

the dregs of the doctoral programs in physics and mathematics." Sagan, *Contact*, 215. In *The Cosmic Connection* Sagan stated much the same sentiment: "There are enormous labor forces and huge electronics, missile, and chemical industries that have an equally strong vested interest in and maintain equally strong lobbies for the maintenance of the warfare state," p. 64.

12. Carl Sagan, "Exotic Biochemistries in Exobiology," in *Extra-Terrestrial Life*, ed. E. Shneour and E. Ottesen (National Academy of Sciences, 1966).

13. Another work of fiction that raised expectations of near planetary unity in the wake of extraterrestrial contact was the 1996 film *Star Trek: First Contact*.

14. Sagan, *Contact*, 146.

15. Sagan's heroine brings to mind the pioneering Australian radio astronomer Ruby Payne-Scott.

16. Sagan, *Contact*, 364.

17. Ibid., 263.

18. Excerpts from copies of the NSF budget sent to Merle Tuve by Raymond Seeger, Acting Assistant Director, NSF, November 10, 1955, Tuve Papers, Library of Congress, box 327, folder AUI-NSF Sept 1955→.

19. My next book exploring normal science in the Cold War will be a study of ecology, botany, and agriculture focusing on the history of phytotrons and controlled environment laboratories. Early work toward that book can be found in "Biological Big Science," a talk given at a Joint Atlantic Seminar in the History of Biology in Philadelphia in 1999. Also see Toby Appel, *Shaping Biology* (Johns Hopkins University Press, 2000), 183–186; Sharon Kingsland, "Frits Went's Atomic Age Greenhouse," *Journal for the History of Biology* 42 (2009), no. 2: 289–324; David P. D. Munns, "Controlling the Environment," *British Scholar* 2 (2010), no. 2: 197–227.

20. The Imperial College Ecotron, however, is a product of the 1990s. Hannah Gay, *The History of Imperial College 1907–2007* (Imperial College Press, 2007), 289, 659.

BIBLIOGRAPHY

MANUSCRIPT SOURCES

AUSTRALIA

Australian Archives, series C3830. Technical and Correspondence Files, CSIR/CSIRO (after 1949) Division of Radiophysics NSW, November 27, 1940–April 1, 1992.

CSIRO Archives, series 3. Head Office Files, 1926–1991 (being reclassified as Australian Archives Series, CRS A8520).

CSIRO Archives, series 4. Head Office Files. Collected Files on Radio Research Board and Radiophysics Advisory Board, no date range.

CSIRO Archives, series 9. Head Office. Correspondence Files, 1926–1952 (being reclassified as Australian Archives Series, CRS A9778).

David Rivett Papers, MS83, Basser Library, Australian Academy of Science.

Joseph L. Pawsey Papers, MS20, Basser Library, Australian Academy of Science.

UNITED STATES

Walter S. Adams Papers, Huntington Library, Pasadena.

Astrophysics Archives, California Institute of Technology Archives.

Walter Baade Papers, Huntington Library, Pasadena.

Bart Bok Papers, Harvard College Archives.

Ira S. Bowen Papers, Huntington Library, Pasadena.

Nan Dieter Papers, Bancroft Library, University of California, Berkeley.

Lee A. DuBridge Papers, California Institute of Technology Archives.

Leo Goldberg Papers, Harvard College Archives.

Jessie Greenstein Papers, California Institute of Technology Archives.

Historical Files, California Institute of Technology Archives.

Edwin Hubble Papers, Huntington Library.

Donald Menzel Papers, Harvard College Archives.

Oral Histories, California Institute of Technology Archives

Owens Valley Papers, California Institute of Technology Archives.

Bruce Rule Papers, California Institute of Technology Archives.

Otto Struve Papers, Bancroft Library, University of California, Berkeley.

Merle Tuve Papers, Library of Congress.

Report of the President of Harvard College and Reports of Departments, 1945–1960, Harvard College Archives.

Series 410D, Record Group 1.2, Rockefeller Foundation Archives, Rockefeller Archive Center, Sleepy Hollow, New York.

Menzel, Donald, unpublished autobiography, ca. 1974, HUG4567.3, Harvard College Archives.

UNITED KINGDOM
Jodrell Bank Archives, John Rylands Library, Manchester.

Martin Ryle Papers, Churchill College, Cambridge University.

Robert Hanbury Brown Papers, Royal Society, London.

Second World War Papers of Sir Bernard Lovell, Imperial War Museum, Netherlands.

Jan Oort Papers, Bibliotheek der Universiteit, Leiden.

ORAL SOURCES

Robert Bacher. Interview with Mary Terrall, 1983. CIT Archives.

Bart J. Bok. *Sources for History of Modern Astrophysics*. Interview with David DeVorkin, May 1978. American Institute of Physics.

John G. Bolton Interview, TRC324, Australian National Library.

Marshall Cohen. Interview with Shelly Irwin, CIT Archives.

Jesse Greenstein, *Sources for History of Modern Astrophysics*. Interview with Spencer Weart, April 7, 1977. American Institute of Physics.

Jesse Greenstein. Interview by Rachel Prud'Homme, 1983. CIT Archives.

Bernard Mills. DeB 584, Australian National Library.

Jan Oort, *Sources for History of Modern Astrophysics*. Interview with David DeVorkin, November 10, 1977. American Institute of Physics.

Sound recording of Sir F. W. G. White being interviewed by David Ellyard, 1979. CSIRO Archives Series 528.

PUBLISHED SOURCES

Abir-Am, Pnina. From Multidisciplinary Collaboration to Transnational Objectivity: International Space as Constitutive of Molecular Biology, 1930–1970. In *Denationalizing Science: The Contexts of International Scientific Practice*, ed. E. Crawford, T. Shinn, and S. Sörlin. Kluwer, 1993.

Abraham, Itty. *The Making of the Indian Atomic Bomb: Science, Secrecy and the Postcolonial State.* Zed Books, 1998.

Agar, Jon. Screening Science: Spatial Organization and Valuation at Jodrell Bank. In *Making Space for Science: Territorial Themes in the Shaping of Knowledge*, ed. C. Smith and J. Agar. Macmillan, 1998.

Agar, Jon. *Science and Spectacle: The Work of Jodrell Bank in Post-War British Culture*. Harwood, 1998.

Akera, Atsushi. *Calculating a Natural World: Scientists, Engineers, and Computers during the Rise of U.S. Cold War Research*. MIT Press, 2007.

Alvarez, Luis W. *Alvarez: Adventures of a Physicist*. Basic Books, 1987.

Anderson, Benedict. *Imagined Communities.* Verso, 2003.

Appel, Toby. *Shaping Biology: The National Science Foundation and American Biological Research, 1945–1975*. Johns Hopkins University Press, 2000.

Baade, W., and R. Minkowski. Identification of the Radio Sources in Cassiopeia, Cygnus A, and Puppis A. *Astrophysical Journal* 119 (1954): 206–214.

Baade, W., and R. Minkowski. On the Identification of the Radio Sources. *Astrophysical Journal* 119 (1954): 214–231.

Babcock, H. W. Ira Sprague Bowen. In *Biographical Memoirs*, volume 53. National Academy Press, 1982.

Baird, David. Analytical Chemistry and the "Big" Scientific Instrumentation Revolution. *Annals of Science* 50 (1993): 267–290.

Bamford, James. *The Puzzle Palace: A Report on America's Most Secret Agency*. Penguin, 1983.

Barnett, Correlli. *The Lost Victory: British Dreams, British Realities, 1945–50*. Macmillan, 1995.

Baxter, James Phinney. *Scientists Against Time*. MIT Press, 1968.

Beer, Arthur, ed. *Vistas in Astronomy*. Pergamon, 1960.

Berkner, L.V. Symposium: Radio Telescopes, Present and Future. *Astronomical Journal* 61 (1956), no. 4: 165–170.

Berkner, Lloyd. *The Scientific Age: The Impact of Science on Society*. Yale University Press, 1964.

Biagioli, Mario. *Galileo, Courtier: The Practice of Science in the Culture of Absolutism*. University of Chicago Press, 1993.

Blackett, P. M. S. *Fear, War and the Bomb*. McGraw-Hill, 1948.

Bok, B. J. Radio Studies of Interstellar Hydrogen. *Sky and Telescope* 13 (1954), October: 408–412.

Bok, B. J. New Science of Radio Astronomy. *Scientific Monthly* 80 (1955): 333–345.

Bok, B. J. Jodrell Bank Symposium on Radio Astronomy. *Sky and Telescope* 15 (1955), November: 21–27.

Bok, B. J. A National Radio Astronomy Observatory. *Scientific American* 195 (1956): 56–64.

Bolton, Geoffrey. *The Oxford History of Australia*, volume 5: *The Middle Way*. Oxford University Press, 1996.

Bolton, J. G. Discrete Sources of Galactic Radio Frequency Noise. *Nature* 162 (1948), July: 141–142.

Bolton, J. G. History of Australian Astronomy: Radio Astronomy at Dover Heights. *Proceedings of the Astronomical Society of Australia* 4 (1982), no. 4: 349–358.

Bowen, E. G. History of Australian Astronomy: The Pre-History of the Parkes 64-m Telescope. *Proceedings of the Astronomical Society of Australia* 4 (1981), no. 2: 267–273.

Bowen, E. G. The Origins of Radio-Astronomy in Australia. In *The Early Years of Radio Astronomy*, ed. W. Sullivan III. Cambridge University Press, 1984.

Bowen, E. G. *Radar Days*. Adam Hilger, 1987.

Bowen, E. G., O. O. Pulley, and J. S. Gooden. Application of Pulse Technique to the Acceleration of Elementary Particles. *Nature* 157 (1946): 840.

Brown, Louis. *A Radar History of World War II: Technical and Military Imperatives*. Institute of Physics Publishing, 1999.

Brown, Nicolas. *Governing Prosperity: Social Change and Social Analysis in Australia in the 1950s*. Cambridge University Press, 1995.

Browne, Janet. *Charles Darwin: The Power of Place*. Jonathan Cape, 2002.

Bud, Robert. *Penicillin: Triumph and Tragedy*. Oxford University Press, 2007.

Buderi, R. *The Invention That Changed the World: How a Small Group of Radar Pioneers Won the Second World War and Launched a Technological Revolution.* Simon & Schuster, 1996.

Bunge, Mario, and William R. Shea, eds. *Rutherford and Physics at the Turn of the Century.* Dawson, 1979.

Burke, Bernard. Radio Astronomy: The Progress of a Technique-Oriented Discipline. In *Serendipitous Discoveries in Radio Astronomy*, ed. K. Kellermann and B. Sheets. National Radio Astronomy Observatory, 1983.

Bush, Vannevar. *Science—The Endless Frontier, A Report to the President.* Government Printing Office, 1945.

Butrica, Andrew J. *To See the Unseen: A History of Planetary Radio Astronomy.* NASA, 1996.

Capshew, James, and Rader, Karen. Big Science: Price to the Present. *Osiris* (second series) 7 (1992): 3–25.

Caro, D. E., and R. L. Martin. Leslie Harold Martin. *Historical Records of Australian Science* 7 (1987), no. 1: 97–107.

Carroll, James. *House of War: The Pentagon and the Disastrous Rise of American Power.* Houghton Mifflin, 2006.

Carson, Cathryn, Ethan Pollock, Peter Westwick, and James H. Williams. Editors' Foreword. *Historical Studies in the Physical and Biological Sciences* 30 (1999), no. 1: iii–ix.

Casey, R. G. *Friends and Neighbours: Australia and the World.* Cheshire, 1954.

Cawte, Alice. *Atomic Australia, 1944–1990.* University of New South Wales Press, 1992.

Chauncey, George. *Gay New York: Gender, Urban Culture, and the Makings of the Gay Male World, 1890–1940.* Basic Books, 1994.

Choi, Hyaeweol. *An International Scientific Community: Asian Scholars in the United States.* Praeger, 1995.

Christiansen, W. N. History of Australian Astronomy: History and Propaganda in Astronomy. *Proceedings of the Astronomical Society of Australia* 8 (1989), no. 1: 96–101.

Christianson, Gale E. *Edwin Hubble: Mariner of the Nebulae.* Farrar, Straus, and Giroux, 1995.

Clark, Ronald. *The Rise of the Boffins.* Phoenix House, 1962.

Clarke, Arthur C. The Astronomer's New Weapons: Electronic Aids to Astronomy. *Journal of the British Astronomical Association* 55 (1945): 143–147.

Clarke, Arthur C. *The Fountains of Paradise.* VGSF, 1989.

Close, Colonel Sir Charles. *The Early Years of the Ordnance Survey.* David & Charles, 1969.

Cockburn, Stewart, and David Ellyard. *Oliphant: The Life and Times of Sir Mark Oliphant.* Axiom Books, 1992.

Cohen, I. Bernard. *Science, Servant of Man: A Layman's Primer for the Age of Science*. Sigma Books, 1949.

Cohen-Cole, Jamie. Instituting the Science of Mind: Intellectual Economics and Disciplinary Exchange at Harvard's Center for Cognitive Studies. *British Journal for the History of Science* 40 (2007): 567–597.

Coombs, H. C. *Trial Balance*. Macmillan, 1981.

Couffer, Jack. *Bat Bomb: World War II's Other Secret Weapon*. University of Texas Press, 1992.

Cowan, Ruth Schwartz. *More Work for Mother: The Ironies of Household Technology from the Open Hearth to the Microwave*. Basic Books, 1983.

Crane, Diana. *Invisible Colleges: Diffusion of Knowledge in Scientific Communities*. University of Chicago Press, 1972.

Crawford, Elisabeth, Terry Shinn, and Sverker Sörlin. The Nationalization and Denationalization of the Science: An introductory essay. In *Denationalizing Science: The Contexts of International Scientific Practice*, ed. E. Crawford, T. Shinn, and S. Sörlin. Kluwer, 1993.

Creager, Angela N. H. Tracing the Politics of Changing Postwar Research Practice: The Export of "American" Radioisotopes to European Biologists. *Studies in the History and Philosophy of Biological and Biomedical Sciences* 33 (2002), no. 3: 367–388.

Crease, Robert. *Making Physics: A Biography of Brookhaven National Laboratory, 1940–1972*. University of Chicago Press, 1999.

Croarken, Mary. The Beginnings of the Manchester Computer Phenomenon: People and Influences. *IEEE Annals of the History of Computing* 15 (1993), no. 3: 9–16.

Darnton, Robert. *The Great Cat Massacre and Other Adventures in French Cultural History*. Basic Books, 1984.

Davies, Paul. *Are We Alone? Implications of the Discovery of Extraterrestrial Life*. Penguin, 1995.

De Maria, Michelangelo, ed. *The Restructuring of Physical Sciences in Europe and the United States, 1945–60*. World Scientific, 1989.

Dennis, Michael Aaron. Accounting for Research: New Histories of Corporate Laboratories and the Social History of American Science. *Social Studies of Science* 17 (1987): 479–518.

Dennis, Michael Aaron. A Change of State: the political cultures of technical practice at the MIT Instrumentation Laboratory and the Johns Hopkins University Applied Physics Laboratory, 1930–45. PhD thesis, Johns Hopkins University, 1990.

Dennis, Michael Aaron. Secrecy and Science Revisited: From Politics to Historical Practice and Back. In *The Historiography of Contemporary Science, Technology, and Medicine: Writing Recent Science*, ed. R. Doel and T. Söderqvist. Routledge, 2006.

DeVorkin, David. Electronics in Astronomy: Early Applications of the Photoelectric Cell and Photomultiplier for Studies of Point-Source Celestial Phenomena. *Proceedings of the IEEE* 73 (1985), no. 3: 1205–1220.

DeVorkin, David. Along for the Ride: The Response of American Astronomers to the Possibility of Space Research, 1945–50. In *The Restructuring of Physical Sciences in Europe and the United States, 1945–60*, ed. M. De Maria. World Scientific, 1989.

DeVorkin, David. *Science with a Vengeance: How the Military Created the US Space Sciences after World War II*. Springer-Verlag, 1992.

DeVorkin, David. Who Speaks for Astronomy? How Astronomers Responded to Government Funding after World War II. *Historical Studies in the Physical and Biological Sciences* 31 (2000), no.1: 52–92.

Doel, Ronald E. *Solar System Astronomy in America: Communities, Patronage, and Interdisciplinary Science*. Cambridge University Press, 1996.

Doel, Ronald E., and Thomas Söderqvist, eds. *The Historiography of Contemporary Science, Technology, and Medicine: Writing Recent Science*. Routledge, 2006.

DuBridge, Lee A. The Effects of World War II on the Science of Physics. *American Journal of Physics* 17 (1949), no. 5: 273–281.

Edge, David, and Michael Mulkay. *Astronomy Transformed: The Emergence of Radio Astronomy in Britain*. Wiley, 1976.

Edgerton, David. 2006. *Warfare State: Britain, 1920–1970*. Cambridge University Press.

Edmondson, Frank K. AURA and KPNO: The Evolution of an Idea, 1952–58. *Journal for the History of Astronomy* 22 (1991): 68–86.

Emberson, Richard M. National Radio Astronomy Observatory: The Early History and Development of the Observatory at Green Bank, West Virginia. *Science* 130 (1956): 1307–1318.

Emberson, Richard M., and N. L. Ashton. The Telescope Program for the National Radio Astronomy Observatory at Green Bank, West Virginia. *Proceedings of the Institute of Radio Engineers* 46 (1958), no. 1: 23–35.

Engel, J. Ronald. *Sacred Sands: The Struggle for Community in the Indiana Dunes*. Wesleyan University Press, 1983.

Ewen, H. I., and E. M. Purcell. Observation of a Line in the Galactic Radio Spectrum— Radiation from Galactic Hydrogen at 1,420 Mcs. *Nature* 168 (1951): 356.

Ewen, H. I. Radiation from Galactic Hydrogen at 1420 Mc/sec. PhD thesis, Harvard University, 1951. HU 90.5951, Harvard College Archives.

Faragher, John Mack. *Sugar Creek: Life on the Illinois Prairie*. Yale University Press, 1986.

Field, Clive, and John Pickstone, eds. *A Centre of Intelligence: The Development of Science, Technology and Medicine in Manchester and Its University*. John Rylands Library, 1988.

Fitzgerald, Frances. *Cities on a Hill: A Journey through Contemporary American Cultures*. Simon & Schuster, 1987.

Forman, Paul. Scientific Internationalism and the Weimar Physicists: The Ideology and Its Manipulation in Germany after World War I. *Isis* 64 (1973): 151–180.

Forman, Paul. Behind Quantum Electronics: National Security as a Basis for Physical Research in the United States, 1940–1960. *Historical Studies in the Physical and Biological Sciences* 18 (1987): 149–229.

Forman, Paul. Inventing the Maser in Postwar America. *Osiris* 7 (1992): 103–134.

Forman, Paul. "Swords into Ploughshares": Breaking New Ground with Radar Hardware and Technique in Physical Research after World War II. *Reviews of Modern Physics* 67 (1995), no. 2: 397–455.

Forman, Paul. Into Quantum Electronics: The Maser as "Gadget" of Cold-War America. In *National Military Establishments and the Advancement of Science and Technology*, ed. P. Forman and J. Sánchez-Ron. Kluwer, 1996.

Fortes, Jacqueline, and Larissa Lomnitz. *Becoming a Scientist in Mexico: The Challenge of Creating a Scientific Community in an Undeveloped Country*. Pennsylvania State University Press, 1994.

Frayn, Michael. Copenhagen. Methuen, 1998.

Freeman, Joan. *A Passion for Physics: The Story of a Woman Physicist*. Adam Hilger, 1991.

Galison, Peter. Bubble Chambers and the Experimental Workplace. In *Observation, Experiment, and Hypothesis in Modern Physical Science*, ed. P. Achinstein and O. Hannaway. Cambridge University Press, 1985.

Galison, Peter. *How Experiments End*. University of Chicago Press, 1987.

Galison, Peter, and Bruce Hevly, eds. *Big Science: The Growth of Large-Scale Research*. Stanford University Press, 1992.

Galison, Peter, Bruce Hevly, and Rebecca Lowen. Controlling the Monster: Stanford and the Growth of Physics Research, 1935–1962. In *Big Science: The Growth of Large-Scale Research*, ed. P. Galison and B. Hevly. Stanford University Press, 1992.

Galison, Peter. *Image and Logic: Towards a Material Culture of Microphysics*. University of Chicago Press, 1997.

Galison, Peter. *Einstein's Clocks, Poincaré's Maps: Empires of Time*. Norton, 2004.

Gaudillière, Jean-Paul. Paris–New York Roundtrip: Transatlantic Crossings and the Reconstruction of the Biological Sciences in Post-War France. *Studies in the History and Philosophy of Biological and Biomedical Sciences* 33 (2002), no. 3: 389–417.

Gay, Hannah. *The History of Imperial College 1907–2007: Higher Education and Research in Science, Technology, and Medicine*. Imperial College Press, 2007.

Geiger, Roger. *Research and Relevant Knowledge: American Research Universities since World War II*. Oxford Press, 1993.

Genuth, Sara Schechner. From Heaven's Alarm to Public Appeal: Comets and the Rise of Astronomy at Harvard. In *Science at Harvard University: Historical Perspectives*, ed. C. Elliot and M. Rossiter. Associated University Presses, 1992.

Gillespie, D. T. C. Research Management in the Commonwealth Scientific and Industrial Research Organization, Australia. *Public Administration* 46 (1964), no. 1: 11–31.

Ginzton, E. L., W. W. Hansen, and W. R. Kennedy. A Linear Electron Accelerator. *Review of Scientific Instruments* 19 (1948): 89–108.

Goldman, Joanne. National Science in the Nation's Heartland: The Ames Laboratory and Iowa State University, 1942–1962. *Technology and Culture* 41 (2000): 432–459.

Goldstein, Jack. *A Different Sort of Time: The Life of Jerrold R. Zacharias*. MIT Press, 1992.

Golinski, Jan. *Making Natural Knowledge: Constructivism and the History of Science*. Cambridge University Press, 1998.

Gooday, Graeme. Precision Measurement and the Genesis of Physics Teaching Laboratories in Victorian Britain. *British Journal for the History of Science* 23 (1990): 25–51.

Goodstein, Judith R. *Millikan's School: A History of the California Institute of Technology*. Norton, 1991.

Gordin, Michael D. *Five Days in August: How World War II Became a Nuclear War*. Princeton University Press, 2007.

Goss, W. M., and R. X. McGee. *Under the Radar: The First Woman Radio Astronomer: Ruby Payne-Scott*. Springer, 2009.

Graham, J. A., C. M. Wade, and R. M. Price. Bart J. Bok, 1906–1983. National Academy of Sciences, 1994.

Graham Smith, F., and Lovell, Bernard. On the Discovery of Extragalactic Radio Sources. *Journal for the History of Astronomy* 14 (1983): 155–165.

Greenstein, J., and F. L. Whipple. On the Origin of the Interstellar Radio Disturbances. *Proceedings of the National Academy of Sciences of the United States of America* 23 (1937): 177.

Greenstein, J., L. Henyey, and P. G. Keenan. Interstellar Origin of Cosmic Radiation at Radio-Frequencies. *Nature* 157 (1946): 805–806.

Greenstein, J. L. Washington Conference on Radio Astronomy—1954. *Journal of Geophysical Research* 59 (1954): 149–201.

Greenstein, Jesse L. An Astronomical Life. *Annual Review of Astronomy and Astrophysics* 22 (1984): 1–35.

Guerlac, Henry. *Radar in World War II.* Tomash and American Institute of Physics, 1987.

Gusfield, Joseph R. *Community: A Critical Response.* Harper & Row, 1971.

Hacking, Ian. Introduction. In *Scientific Practice: Theories and Stories of Doing Physics*, ed. J. Buchwald. University of Chicago Press, 1995.

Hagan, J. Radio Astronomy Conference. *Science* 119 (1954): 588–591.

Hagen, Joel B. *An Entangled Bank: The Origins of Ecosystem Ecology.* Rutgers University Press, 1992.

Hanbury Brown, R. *Boffin: A Personal Story of the Early Days of Radar, Radio Astronomy and Quantum Optics.* Adam Hilger, 1991.

Hanbury Brown, R., H. C. Minnett, and F. W. G. White. Edward George Bowen. *Biographical Memoirs of Fellows of the Royal Society* 38 (1992): 43–65.

Hart, David M. *Forged Consensus: Science, Technology, and Economic Policy in the United States, 1921–1953.* Princeton University Press, 1998.

Harwood, Jonathan. *Styles of Scientific Thought: The German Genetics Community, 1900–1933.* University of Chicago Press, 1993.

Hård, Mikael. Technological Drift in Science: Swedish Radio Astronomy in the Making, 1942–1972. In *Center on the Periphery: Historical Aspects of Twentieth Century Swedish Physics*, ed. S. Lindqvist. Science History, 1993.

Haynes, R. From Swords to Ploughshares. In *Explorers of the Southern Sky: A History of Australian Astronomy*, ed. R. Haynes, R. Haynes, D. Malin, and R. McGee. Cambridge University Press, 1996.

Haynes, R., R. Haynes, D. Malin, and R. McGee, eds. *Explorers of the Southern Sky: A History of Australian Astronomy.* Cambridge University Press, 1996.

Heeschen, David S. Harvard's New Radio Telescope. *Sky and Telescope* 15 (1956), July: 388–390.

Heilbron, John, and Robert Seidel. *Lawrence and His Laboratory: A History of the Lawrence Berkeley Laboratory.* University of California Press, 1989.

Heims, Steve Joshua. *Constructing a Social Science for Postwar American: The Cybernetics Group, 1946–1953.* MIT Press, 1993.

Hensley, Oliver D. Traditional Models for Classifying University Research. In *The Classification of Research*, ed. O. Hensley. Texas Tech University Press, 1988.

Hermann, Armin, John Krige, Ulrike Mersits, and Dominique Pestre. *History of CERN*. North-Holland, 1987.

Hevly, Bruce W. Basic Research within a Military Context: The Naval Research Laboratory and the Foundations of Extreme Ultraviolet and X-Ray Astronomy. PhD thesis, Johns Hopkins University, 1987.

Hevly, B. Reflections on Big Science and Big History. In *Big Science: The Growth of Large-Scale Research*, ed. P. Galison and B. Hevly. Stanford University Press, 1992.

Hey, J. S. Radio Astronomy. *Science Progress* 39 (1951): 427–448.

Hey, J. S. *The Evolution of Radio Astronomy*. Science History Publications, 1973.

Hirsh, Richard F. *Glimpsing an Invisible Universe: The Emergence of X-Ray Astronomy*. Cambridge University Press, 1983.

Hoffleit, Dorrit. New Notes. *Sky and Telescope* 10 (1951), November: 10.

Home, R. W. Science on Service, 1939–45. In Australian Science in the Making, ed. R. Home. Cambridge University Press, 1988.

Home, R. W. The Rush to Accelerate: Early Stages in Nuclear Physics Research in Australia. *Historical Studies in the Physical and Biological Sciences* 36 (2006), no. 2: 213–241.

Hounshell, David. *From the American System to Mass Production, 1800–1932: The Development of Manufacturing Technology in the United States*. Johns Hopkins University Press, 1985.

Hufbauer, K. *Exploring the Sun: Solar Science since Galileo*. Johns Hopkins University Press, 1991.

Hunt, B. J. Michael Faraday, Cable Telegraphy and the Rise of Field Theory. *History and Technology* 13 (1991): 1–19.

Iliffe, Rob. Material Doubts: Hooke, Artisan Culture and the Exchange of Information in 1670s London. *British Journal for the History of Science* 28 (1995): 285–318.

International Astronomical Union and International Scientific Radio Union. *Paris Symposium on Radio Astronomy: IAU Symposium No. 9 and URSI Symposium No. 1*, ed. R. Bracewell. Stanford University Press, 1959.

Jackson, Catherine. Visible Work: The Role of Students in the Creation of Liebig's Giesson Research School. *Notes and Records of the Royal Society* 62 (2008): 31–49.

Jacobs, Struan. Scientific Community: Formulations and Critique of a Sociological Motif. *British Journal of Sociology* 38 (1987): 266–276.

Jardine, Lisa. *Erasmus, Man of Letters: The Construction of Charisma in Print*. Princeton University Press, 1993.

Jardine, Lisa. *The Curious Life of Robert Hooke: The Man Who Measured London*. Harper, 2003.

Johnson, Ann. Modeling Molecules: Computational Nanotechnology as a Knowledge Community. *Perspectives on Science* 17 (2009): 144–173.

Jones, Bessie, and Lyle Boyd. *The Harvard College Observatory: The First Four Directorships, 1839–1919.* Belknap, 1971.

Kaiser, David. "A Ψ is just a Ψ?" Pedagogy, Practice, and the Reconstitution of General Relativity, 1942–1975. *Studies in the History and Philosophy of Modern Physics* 29 (1998), no. 3: 321–338.

Kaiser, David I. Making Theory: Producing Physics and Physicists in Postwar America. PhD thesis, Harvard University, 2000.

Kaiser, David. Cold War Requisitions, Scientific Manpower, and the Production of American Physicists after World War II. *Historical Studies in the Physical and Biological Sciences* 33 (2002), no. 1: 131–159.

Kaiser, David. The Postwar Suburbanization of American Physics. *American Quarterly* 56 (2004): 851–888.

Kaiser, David. Moving Pedagogy from the Periphery to the Center. In Pedagogy and the Practice of Science: Historical and Contemporary Perspectives, ed. D. Kaiser and A. Warwick. MIT Press, 2005.

Kaiser, David, and Andrew Warwick, eds. *Pedagogy and the Practice of Science: Historical and Contemporary Perspectives.* MIT Press, 2005.

Kargon, Robert H. *Science in Victorian Manchester.* Manchester University Press, 1977.

Kargon, Robert H. Temple to Science: Cooperative Research and the Birth of the California Institute of Technology. *Historical Studies in the Physical Sciences* 8 (1977): 3–31.

Kargon, Robert H., and Arthur P. Molella. *Artificial Edens: Techno-Cities of the Twentieth Century.* MIT Press, 2008.

Kay, Lily E. *The Molecular Vision of Life.* Oxford University Press, 1993.

Keller, Geoffrey. Report of the Advisory Panel on Radio Astronomy. *Astrophysical Journal* 134 (1961): 927–939.

Keller, Morton, and Phyllis Keller. *Making Harvard Modern: The Rise of America's University.* Oxford University Press, 2001.

Keller, S. The American Dream of Community: An Unfinished Agenda. *Sociological Forum* 3 (1988): 167–183.

Kellermann, K. John Gatenby Bolton (1922–1993). *Publications of the Astronomical Society of the Pacific* 108 (1996): 729–737.

Kellermann, K., and B. Sheets, eds. *Serendipitous Discoveries in Radio Astronomy: Proceedings of a Workshop Held at the National Radio Astronomy Observatory.* National Radio Astronomy Observatory, 1983.

Kerr, Frank. Early Days in Radio and Radar Astronomy in Australia. In *The Early Years of Radio Astronomy*, ed. W. Sullivan III. Cambridge University Press, 1984.

Kerr, Frank. Radio Astronomy at the URSI Assembly. *Sky and Telescope* 12 (1953): 59–62, 70.

Kevles, Daniel. Jodrell Bank and Cambridge. *Science* 196 (1977): 774–776.

Kevles, Daniel J. Principles and Politics in Federal R&D Policy, 1945–90: An Appreciation of the Bush Report. Preface to 1990 edition of *Science—The Endless Frontier*. National Science Foundation, 1990.

Kevles, Daniel J. *The Physicists: The History of a Scientific Community in Modern America*. Harvard University Press, 1995.

Kevles, Daniel. Big Science and Big Politics in the United States: Reflection on the Death of the SSC and the Life of the Human Genome Project. *Historical Studies in the Physical and Biological Sciences* 27 (1997), no. 2: 269–298.

Kidwell, Peggy. E. C. Pickering, Lydia Hinchman, Harlow Shapley, and the Beginning of Graduate Work at the Harvard College Observatory. *Astronomy Quarterly* 5 (1986): 157–171.

Kidwell, Peggy A. Cecilia Payne-Gaposchkin: Astronomy in the Family. In Uneasy Careers and Intimate Lives: Women in Science, 1789–1979, ed. P. Abir-Am and D. Outram. Rutgers University Press, 1987.

Kidwell, Peggy. Harvard Astronomers and World War II: Disruption and Opportunity. In *Science at Harvard University: Historical Perspectives*, ed. C. Elliot and M. Rossiter. Associated University Presses, 1992.

Kirsch, Scott. *Proving Grounds: Project Plowshare and the Unrealized Dream of Nuclear Earthmoving*. Rutgers University Press, 2005.

Kleinman, Daniel. *Politics on the Endless Frontier: Postwar Research Policy in the United States*. Duke University Press, 1995.

Kohler, Robert E. The PhD Machine: Building on the Collegiate Base. *Isis* 81 (1990): 639–662.

Kohler, Robert E. *Partners in Science: Foundations and Natural Scientists, 1900–1945*. University of Chicago Press, 1991.

Kohler, Robert E. *Lords of the Fly*. University of Chicago Press, 1994.

Kohlstedt, Sally G. *The Formation of the American Scientific Community: The American Association for the Advancement of Science, 1848–60*. University of Illinois, 1976.

Kolb, A., and L. Hoddeson. The Mirage of the "World Accelerator for World Peace" and the Origin of the SSC, 1953–1983. *Historical Studies in the Physical and Biological Sciences* 24 (1993), no. 1: 101–124.

Kopal, Zdenêk. *Of Stars and Men: Reminiscences of an Astronomer*. Adam Hilger, 1986.

Kragh, Helge. *Dirac: A Scientific Biography*. Cambridge University Press, 1990.

Krige, J. The Installation of High-Energy Accelerators in Britain after the War: Big Equipment but Not Big Science. In *The Restructuring of Physical Science in Europe and the United States, 1945–60*, ed. M. De Maria. World Scientific, 1989.

Krige, John. *American Hegemony and the Postwar Reconstruction of Science in Europe*. MIT Press, 2006.

Kuhn, Thomas S. *The Structure of Scientific Revolutions*. University of Chicago Press, 1996.

Lankford, John. *American Astronomy: Community, Careers, and Power, 1859–1940*. University of Chicago Press, 1997.

Latour, Bruno, and Steve Woolgar. *Laboratory Life: The Construction of Scientific Facts*. Princeton University Press, 1986.

LeCarré, John. *The Spy Who Came In From The Cold*. Scribner, 2001.

Lemaine, G., R. MacLeod, M. Mulkay, and P. Weingart. 1976. Introduction: Problems in the Emergence of New Disciplines. In *Perspectives on the Emergence of Scientific Disciplines*, ed. G. Lemaine et al. De Gruyter Mouton.

Lenoir, Timothy, and Christophe Lécuyer. Instrument Makers and Discipline Builders: The Case of Nuclear Magnetic Resonance. *Perspectives on Science* 3 (1995): 276–345.

Leslie, Stuart W. Playing the Education Game to Win: The Military and Interdisciplinary Research at Stanford. *Historical Studies in the Physical and Biological Sciences* 18 (1987): 55–88.

Leslie, Stuart. *The Cold War and American Science: The Military-Industrial-Academic Complex at MIT and Stanford*. Columbia University Press, 1993.

Leslie, Stuart, and Bruce Hevly. Steeple Building at Stanford: Electrical Engineering, Physics, and Microwave Research. *IEEE Proceedings* 10 (1985): 1161–1180.

Levy, D. H. *The Man Who Sold the Milky Way: A Biography of Bart Bok*. University of Arizona Press, 1993.

Long, Pamela O. *Openness, Secrecy, Authorship: Technical Arts and the Culture of Knowledge from Antiquity to the Renaissance*. Johns Hopkins University Press, 2004.

Lovell, A. C. B. Radio Astronomy at Jodrell Bank—I. *Sky and Telescope* 12 (1953), February: 94–96, 114.

Lovell, A. C. B. Robert Hanbury Brown. In *Modern Technology and Its Influence on Astronomy*, ed. J. Wall and A. Boksenberg. Cambridge University Press, 1990.

Lovell, B. Joseph Lade Pawsey. *Biographical Memoirs of Fellows of the Royal Society* 10 (1964): 229–243.

Lovell, B. *The Story of Jodrell Bank*. Harper & Row, 1968.

Lovell, B. *The Jodrell Bank Telescopes*. Oxford University Press, 1985.

Lovell, B. *Astronomer by Chance*. Basic Books, 1990.

Lovell, B. *Echoes of War: The Story of H2S Radar*. Adam Hilger, 1991.

Lovell, B., and J. A. Clegg. *Radio Astronomy*. Wiley, 1952.

Lowen, Rebecca. *Creating the Cold War University: The Transformation of Stanford*. University of California Press, 1997.

Mahoney, Michael S. Computer Science: The Search for a Mathematical Theory. In *Companion to Science in the Twentieth Century*, ed. J. Krige and D. Pestre. Routledge, 2003.

Malphrus, Benjamin. *The History of Radio Astronomy and the National Radio Astronomy Observatory: evolution toward big science*. Krieger, 1996.

Manzione, Joseph. The American Scientific Community, the United States Government, and the Issue of International Scientific Relations during the Cold War, 1945–1960. PhD thesis, University of Michigan, 1992.

May, Elaine Tyler. *Homeward Bound: American Families in the Cold War*. Basic Books, 1999.

McCray, W. Patrick. *Giant Telescopes: Astronomical Ambition and the Promise of Technology*. Harvard University Press, 2006.

McDougall, Walter. . . . *The Heavens and the Earth: A Political History of the Space Race*. Basic Books, 1985.

Mellor, D. P. *The Role of Science and Industry*. Australian War Memorial, 1958.

Mills, B. Y., and A. Little. A High-Resolution Aerial System of a New Type. *Australian Journal of Physics* 6 (1953): 272–278.

Minnett, H. C., and Rutherford Robertson. Frederick William George White. *Historical Records of Australian Science* 11 (1996), no. 2: 239–258.

Mitter, Rana, and Patrick Major. Foreword. *Cold War History* 4 (2003).

Munns, David P. D. Linear Accelerators, Radio Astronomy, and Australia's Search for International Prestige, 1944–1948. *Historical Studies in the Physical and Biological Sciences* 27 (1997), no. 2: 299–317.

Munns, David P. D. If We Build It, Who Will Come? Training Radio Astronomers and the Limitations of "National" Laboratories in Cold War America. *Historical Studies in the Physical and Biological Sciences* 34 (2003), no. 1: 93–117.

Needell, Allan. Nuclear Reactors and the Founding of Brookhaven National Laboratory. *Historical Studies in the Physical Sciences* 14 (1984): 93–122.

Needell, Allan A. Preparing for the Space Age: University-Based Research, 1946–57. *Historical Studies in the Physical and Biological Sciences* 18 (1987): 89–109.

Needell, Allan. Lloyd Berkner, Merle Tuve, and the Federal Role in Radio Astronomy. *Osiris* 3 (1987): 261–288.

Needell, Allan. The Carnegie Institution of Washington and Radio Astronomy: Prelude to an American National Observatory. *Journal for the History of Astronomy* 22 (1991): 55–67.

Needell, Allan A. *Science, Cold War and the American State: Lloyd V. Berkner and the Balance of Professional Ideals*. Harwood, 2000.

Nye, David E. *American Technological Sublime*. MIT Press, 1999.

Nye, Mary Jo. *Before Big Science: The Pursuit of Modern Chemistry and Physics, 1800–1940*. Prentice Hall, 1996.

Nye, Mary Jo. *Blackett: Physics, War, and politics in the Twentieth Century*. Harvard University Press, 2004.

Olesko, Kathryn M. *Physics as a Calling: Discipline and Practice in the Königsberg Seminar for Physics*. Cornell University Press, 1991.

Osterbrock, Donald. The Appointment of a Physicist as Director to the Astronomical Center of the World. *Journal for the History of Astronomy* 23 (1992): 155–165.

Osterbrock, Donald. *Pauper and Prince: Ritchey, Hale, and Big American Telescopes*. University of Arizona Press, 1993.

Osterbrock, Donald. *Yerkes Observatory, 1892–1950: The Birth, Near Death, and Resurrection of a Scientific Research Institution*. University of Chicago Press, 1997.

Overbye, Dennis. *Lonely Hearts of the Cosmos: The Scientific Quest for the Secret of the Universe*. HarperCollins, 1991.

Palmer, H. P. International Conference of Radio Astronomers. *Nature* 184 (1959): 1755–1756.

Pang, Alex Soojung-Kim. Visual Representation and Post-Constructivist History of Science. *Historical Studies in the Physical and Biological Sciences* 27 (1997): 139–171.

Pawsey, J. L. Atomic Power and American Work on the Development of the Atomic Bomb. *Australian Journal of Science* 8 (1945), no. 1: 41–47.

Pawsey, J. L. The Use of Radio Waves for Astronomical Observations. *Australian Journal of Science* 12 (1949), no.1: 5–12.

Pawsey, J. L. Radio Astronomy in Australia. *Journal of the Royal Astronomical Society of Canada* 47 (1953): 137–152.

Pawsey, J. L. Australian Radio Astronomy: How It Developed in This Country. *Australian Scientist* 1 (1961), April: 181–186.

Pawsey, J., R. Payne-Scott, and L. McCready. Radio-Frequency Energy from the Sun. *Nature* 157 (1946): 158–159.

Perrin, Noel. *Giving Up the Gun: Japan's Reversion to the Sword, 1543–1897*. David R. Godine, 2004.

Pestré, Dominique. Studies of the Ionosphere and Forecasts for Radiocommunications. Physicists and Engineer, the Military and National Laboratories in France (and Germany) after 1945. *History and Technology* 13 (1997): 183–205.

Pestré, Dominique. Commemorative Practices at CERN. In *Commemorative Practices in Science: Historical Perspectives on the Politics of Collective Memory*, ed. P. Abir-Am and C. Elliott. *Osiris* (second series) 14 (1999): 215–239.

Pickering, Andrew. Big Science as a Form of Life. In *The Restructuring of Physical Sciences in Europe and the United States, 1945–60*, ed. M. De Maria. World Scientific, 1989.

Primrose, Xanthe. A Powerful Paradox: Australian Public Opinion and Atomic Energy, 1945–1960. BA thesis. Australian National University, 1994.

Purcell, Edward. Observation of a Line in the Galactic Radio Spectrum. *Nature* 168 (1951): 356.

Purcell, Edward. Line Spectra in Radio Astronomy. *Proceedings of the American Academy of Arts and Sciences* 82 (1953), no. 7: 347–349.

Putnam, Robert D. *Bowling Alone: The Collapse and Revival of American Community*. Simon & Schuster, 2000.

Quandt, Jean B. *From the Small Town to the Great Community: The Social Thought of Progressive Intellectuals*. Rutgers University Press, 1970.

Rasmussen, Nicolas. *Picture Control: The Electron Microscope and the Transformation of Biology in America, 1940–60*. Stanford University Press, 1997.

Rasmussen, Nicolas. The Mid-Century Biophysics Bubble: Hiroshima and the Biological Revolution in America, Revisited. *History of Science* 35 (1997): 245–293.

Ratcliffe, J. A. Friday, Nov 23, 1951, Radio Astronomy. *Proceedings of the Royal Institution of Great Britain* 35 (1951): 1–7.

Reber, G., and J. L. Greenstein. Radio-Frequency Investigations of Astronomical Interest. *Observatory* 67 (1947): 15–26.

Reber, Grote. Early Radio Astronomy at Wheaton, Illinois. In *The Early Years of Radio Astronomy*, ed. W. Sullivan III. Cambridge University Press, 1984.

Redfield, Peter. *Space in the Tropics: From Convicts to Rockets in French Guiana*. University of California Press, 2000.

Reeves, Eileen. *Painting the Heavens: Art and Science in the Age of Galileo*. Princeton University Press, 1997.

Reingold, N. Choosing the Future: The US Research Community, 1944–1946. *Historical Studies in the Physical and Biological Sciences* 25 (1995), no. 2: 301–328.

Reynolds, Terry, and Theodore Bernstein. Edison and "The Chair." *IEEE Technology and Society Magazine* 8 (1989), March: 19–28.

Rhodes, Richard. *The Making of the Atomic Bomb*. Penguin Books, 1988.

Robertson, Peter. John Bolton and Australian Astronomy. *Australian Physicist* 21 (1984), no. 8: 178–180.

Robertson, P. *Beyond Southern Skies: Radio Astronomy and the Parkes Telescope*. Cambridge University Press, 1992.

Rosental, Claude. Fuzzyfying the World: Social Practices of Showing the Properties of Fuzzy Logic. In *Growing Explanations: Historical Perspective on Recent Science*, ed. M. Wise. Duke University Press, 2004.

Rothenberg, Marc. Organization and Control: Professionals and Amateurs in American Astronomy, 1899–1918. *Social Studies of Science* 11 (1981): 305–325.

Rowe, A. P. *One Story of Radar*. Cambridge University Press, 1948.

Rowse, T. *Australian Liberalism and National Character*. Kibble Books, 1978.

Ryle, Martin, and Anthony Hewish. The Effects of the Terrestrial Ionosphere on the Radio Waves from Discrete Sources in the Galaxy. *Monthly Notices of the Royal Astronomical Society* 110 (1950): 381.

Ryle, M. Radio Astronomy. *Reports on Progress in Physics* 13 (1950): 184–246.

Ryle, M., and J. A. Ratcliffe. Radio Astronomy. *Endeavor* 11 (1952): 117–125.

Ryle, M. The Mullard Radio Astronomy Observatory, Cambridge. *Nature* 180 (1957), July: 110–112.

Ryle, M. Title to Be Inserted in Proofs. *Observatory* 78 (1958): 61–63.

Ryle, Sir Martin. Radio Astronomy: The Cambridge Contribution. In *Search and Research*, ed. J. Wilson. Mullard, 1971.

Sagan, Carl. Exotic Biochemistries in Exobiology. In *Extra-Terrestrial Life: An Anthology and Bibliography*, ed. E. Shneour and E. Ottesen. National Academy of Sciences, 1966.

Sagan, Carl, ed. *Communication with Extraterrestrial Intelligence (CETI)*. MIT Press, 1973.

Sagan, Carl. *Other Worlds*. Bantam, 1975.

Sagan, Carl. *The Cosmic Connection: An Extraterrestrial Perspective*. Papermac, 1981.

Sagan, Carl. *Contact*. Legend, 1988.

Sagan, Carl. *Pale Blue Dot: A Vision of the Human Future in Space*. Ballantine Books, 1997.

Sagan, Carl. *Carl Sagan's Cosmic Connection: An Extraterrestrial Perspective*, ed. J. Agel. Cambridge University Press, 2000.

Sapolsky, Harvey M. *Science and the Navy: The History of the Office of Naval Research*. Princeton University Press, 1990.

Saward, Dudley. *Bernard Lovell: A Biography*. Robert Hale, 1984.

Schedvin, C. Boris. *Shaping Science and Industry: A History of Australia's Council for Scientific and Industrial Research, 1926–49*. Allen & Unwin, 1987.

Schwartzman, Simon. *A Space for Science: The Development of the Scientific Community in Brazil*. Pennsylvania State University Press, 1991.

Schweber, S. S. The Mutual Embrace of Science and the Military ONR: The Growth of Physics in the United States after World War II, In *Science, Technology, and the Military*, ed. E. Mendelson, M. Smith, and P. Weingart. Kluwer, 1988.

Seidel, Robert. Accelerating Science: The Postwar Transformation of the Lawrence Radiation Laboratory. *Historical Studies in the Physical Sciences* 13 (1983): 375–400.

Servos, John W. The Knowledge Corporation: A. A. Noyes and Chemistry at Caltech, 1915–1930. *Ambix* 23 (1976): 175–186.

Servos, John. *Physical Chemistry from Ostwald to Pauling: The Making of a Science in America*. Princeton University Press, 1990.

Shapin, Steven, and Simon Shaffer. *Leviathan and the Air Pump: Hobbes, Boyle, and the Experimental Life*. Princeton University Press, 1989.

Shapley, Harlow. Report: The Harvard College Observatory. *Science* 101 (1945): 304–305.

Smith, F. G. Apparent Angular Sizes of Discrete Radio Sources. *Nature* 170 (1952), December: 1065.

Smith, F. G. The Measurement of the Angular Diameter of Radio Stars. *Proceedings of the Physical Society B* 65 (1952): 971–980.

Smith, Robert. The Biggest Kind of Big Science: Astronomy and the Space Telescope. In *Big Science: The Growth of Large-Scale Research*, ed. P. Galison and B. Hevly. Stanford University Press, 1992.

Smith, Robert W. *The Space Telescope: A Study of NASA, Science, Technology, and Politics*. Cambridge University Press, 1993.

Smith, Robert W. Engines of Discovery: Scientific Instruments and the History of Astronomy and Planetary Science in the United States in the Twentieth Century. *Journal for the History of Astronomy* 28 (1997): 49–77.

Smith, Robert W. Beyond the Big Galaxy: The Structure of the Stellar System 1900–1952. *Journal for the History of Astronomy* 37 (2006): 307–342.

Smith, Robert, and Joseph Tatarewicz. Replacing a Technology: The Large Space Telescopes and CCDs. *Proceedings of the IEEE* 7 (1985), no. 7: 1221–1235.

Söderqvist, Thomas, ed. *The Historiography of Contemporary Science and Technology*. Harwood, 1997).

Spinardi, Graham. Science, Technology, and the Cold War: The Military Uses of the Jodrell Bank Radio Telescope. *Cold War History* 6 (2006), no. 3: 279–300.

Spufford, Francis. *Backroom Boys: The Secret Return of the British Boffins*. Faber & Faber, 2003.

Stanley, Matthew. *Practical Mystic: Religion, Science, and A. S. Eddington*. University of Chicago Press, 2007.

Struve, O. Progress in Radio Astronomy—II. *Sky and Telescope* 9 (1950): 55–56.

Sullivan, Woodruff, III, ed. *The Early Years of Radio Astronomy*. Cambridge University Press, 1984.

Sullivan, Woodruff, III. Early Years of Australian Radio Astronomy. In *Australian Science in the Making*, ed. R. Home. Cambridge University Press, 1988.

Sullivan, Woodruff, III. 2000. Kapteyn's Influence on the Style and Content of Twentieth Century Dutch Astronomy. In *The Legacy of J. C. Kapteyn: Studies on Kapteyn and the Development of Modern Astronomy*, ed. P. van der Kruit and K. van Berkel. Kluwer, 2000.

Sullivan, Woodruff, III. *Cosmic Noise: A History of Early Radio Astronomy*. Cambridge University Press, 2009.

Toulmin, Stephen. A Historical Reappraisal. In *The Classification of Research*, ed. O. Hensley. Texas Tech University Press, 1988.

Traweek, Sharon. *Beamtimes and Lifetimes: The World of High Energy Physics*. Harvard University Press, 1988.

Tuve, Merle. Technology and National Research Policy. *Physics Today* 7 (1954), no. 1: 6–9.

van de Hulst, H. C. *A Course in Radio Astronomy*. Mimeographed, Leiden, 1951.

van de Hulst, H. C., ed. *Radio Astronomy: International Astronomical Union Symposium No. 4 Held at the Jodrell Bank Experimental Station Near Manchester*. Cambridge University Press, 1957.

Van Helden, Albert, and Hankins, Thomas. Introduction: Instruments in the History of Science. *Osiris* (second series) 9 (1994): 1–7.

van Woerden, Hugo, Willem Brouw, and Henk van de Hulst, eds. *Oort and the Universe: A Sketch of Oort's Research and Person*. Reidel, 1980.

Volti, Rudi. Why Internal Combustion? *American Heritage of Invention and Technology* 6 (1990), no. 2: 42–48.

Wark, I. Physical Sciences. In *Science in Australia: Proceedings of a Seminar Organised by the ANU on the Occasion of the Jubilee of the Commonwealth of Australia* (Cheshire, 1952).

Warwick, Andrew. *Masters of Theory: Cambridge and the Rise of Mathematical Physics*. Chicago University Press, 2003.

Waterman, Alan. Basic Research in the United States. In *Symposium on Basic Research*. American Association for the Advancement of Science, 1959.

Watson-Watt, Robert. *Three Steps to Victory*. Odhams, 1957.

Watson-Watt, Robert. *Man's Means to His End*. Heinemann, 1962.

Weinberg, A. Impact of Large-Scale Science on the United States. *Science* 134 (1961): 161–164.

Weisz, George. *Divide and Conquer: A Comparative History of Medical Specialization*. Oxford University Press, 2006.

Westad, Odd Arne. *The Global Cold War: Third World Interventions and the Making of Our Times*. Cambridge University Press, 2007.

Westfold, K. C. *A Course in Radio Astronomy*. California Institute of Technology, 1959.

Westwick, Peter. *The National Labs: Science in an American System*. Harvard University Press, 2003.

Westwick, Peter. *Into the Black: JPL and the American Space Program, 1976–2004*. Yale University Press, 2007.

Wild, J. P. The Beginnings of Radio Astronomy in Australia. *Records of the Australian Academy of Science* 2, no. 3 (1972): 52–61.

Wilson, R. Jackson. *In Quest of Community: Social Philosophy in the United States, 1860–1920*. Wiley, 1968.

Wise, M. N., ed. *Growing Explanations: Historical Perspectives on Recent Science*. Duke University Press, 2004.

Wisnioski, Matthew H. *Engineers for Change: America's Culture Wars and the New Meanings of Technology, 1964–1974*. MIT Press, 2012.

Whitney, Charles A. *The Discovery of Our Galaxy*. Knopf, 1971.

Woolley, Richard v. d. R. Galactic Noise. *Monthly Notices of the Royal Astronomical Society* 107 (1947): 308–315.

Wright, Helen. *Explorer of the Universe: A Biography of George Ellery Hale*. American Institute of Physics, 1994.

Zachary, G. Pascal. *Endless Frontier: Vannevar Bush, Engineer of the American Century*. Free Press, 1997.

INDEX